D0086814

The Urinary System

Other titles in
Human Body Systems

The Circulatory System by Leslie Mertz

The Digestive System by Michael Windelspecht

The Endocrine System by Stephanie Watson and Kelli Miller

The Lymphatic System by Julie McDowell and Michael Windelspecht

The Muscular System by Amy Adams

The Nervous System and Sense Organs by Julie McDowell

The Reproductive System by Kathryn H. Hollen

The Respiratory System by David Petechuk

The Skeletal System by Evelyn Kelly

The Urinary System

Stephanie Watson

HUMAN BODY SYSTEMS
Michael Windelspecht, Series Editor

Greenwood Press
Westport, Connecticut • London

Library of Congress Cataloging-in-Publication Data

Watson, Stephanie.
The urinary system / Stephanie Watson.
p. cm. — (Human body systems)
Includes bibliographical references and index.
ISBN 0–313–32402–6 (alk. paper)
1. Urinary organs—Diseases—Popular works. 2. Urinary organs—Physiology—Popular works. 3. Urology—Popular works. I. Title. II. Human body systems.
RC900.W38 2004
616.6—dc22 2003067648

British Library Cataloguing in Publication Data is available.

Library of Congress Catalog Card Number: 2003067648
ISBN: 0–313–32402–6

First published in 2004

Greenwood Press, 88 Post Road West, Westport, CT 06881
An imprint of Greenwood Publishing Group, Inc.
www.greenwood.com

Printed in the United States of America

The paper used in this book complies with the Permanent Paper Standard issued by the National Information Standards Organization (Z39.48–1984).

10 9 8 7 6 5 4 3 2 1

Illustrations, unless otherwise credited, are by Sandy Windelspecht.

The *Human Body Systems* series is a reference, not a medical or diagnostic manual. No portion of this series is intended to supplement or substitute medical attention and advice. Readers are advised to consult a physician before making decisions related to their diagnosis or treatment.

Contents

Color photos follow p. 110.

Series Foreword

Human Body Systems is a ten-volume series that explores the physiology, history, and diseases of the major organ systems of humans. An organ system is defined as a group of organs that physiologically function together to conduct an activity for the body. In this series we identify ten major functions. These are listed in Table F.1, along with the name of the organ system responsible for the activity. It is sometimes difficult to specifically define an organ system, because many of our organs have dual functions. For example, the liver interacts with both circulatory and digestive systems, the hypothalamus acts as a junction between the nervous and endocrine systems, and the pancreas has both digestive and endocrine secretions. This complex interaction of organs and tissues in the human body is still not completely understood.

This series is unique in that it provides a one-stop reference source for anyone with an interest in the human body. Whereas other references frequently cover one aspect of human biology, from anatomy and physiology to the prevention of diseases, this series takes a more holistic approach. Each volume not only includes a physiological description of how the system works from the cellular level upward, but also a historical summary of how research on the system has changed since the time of the ancients. This is an important aspect of the series, and one that is frequently overlooked in modern textbooks. In order to understand the successes and problems of modern medicine, it is first important to recognize not only the achievements of the past but also the misunderstandings and challenges of the pioneers in medical research.

For example, a visit to any major educational institution reveals large lecture halls, where science instructors present material to the students on the

TABLE F.1. Organ Systems of the Human Body

Organ System	General Function	Examples
Circulatory	Movement of chemicals through the body	Heart
Digestive	Supply of nutrients to the body	Stomach, small intestine
Endocrine	Maintenance of internal environmental conditions	Thyroid
Lymphatic	Immune system, transport, return of fluids	Spleen
Muscular	Movement	Cardiac muscle, skeletal muscle
Nervous	Processing of incoming stimuli and coordination of activity	Brain, spinal cord
Reproductive	Production of offspring	Testes, ovaries
Respiratory	Gas exchange	Lungs
Skeletal	Support, storage of nutrients	Bones, ligaments
Urinary	Removal of waste products	Bladder, kidneys

anatomy and physiology of the human body. Sometimes these classes include laboratory sessions, but in the study of human biology, especially for students who are not bound for professional schools in medicine, the student's exposure to human biology typically centers on a two-dimensional graphic. Most educators accept this process as a necessary evil of the educational system, but few recognize that, in fact, the large lecture classroom is the product of a change in Egyptian religious beliefs before the start of the current era. During the decline of the Egyptian empires and the simultaneous rise of the ancient Greek culture, the Egyptian religious organizations began to forbid the dissection of the human body. This had a twofold influence on medicine. First, the ending of human dissections meant that medical professionals required lectures from educators, instead of participation in laboratory-based education, which led to the birth of the lecture hall. The second consequence would plague modern medicine for a thousand years. Stripped of their access to human cadavers, researchers studied other "lesser" animals and extrapolated their findings to humans. The practices of the ancient Greeks were passed on over the ages and became the basis for the study of modern medicine. These traditions continue to this day throughout the educational institutions of the world.

The history of human biology parallels the development of modern science. In the seventeenth century, William Harvey's study of blood circulation challenged the long-standing belief of the ancient Greeks that blood was produced in the liver and consumed in the tissues of the body. Harvey's pioneering experimental work had a strong influence on others, and within a century the legacy of the ancient Greeks had collapsed. In the eighteenth century a group of chemists who focused on the chemical reactions of the human body, called the iatrochemists, began to apply chemical laws to human physiology. They were joined by the iatrophysicists, who believed that the human body must operate under the physical laws of the universe. This in turn led to the beginnings of organic chemistry and biochemistry in the nineteenth century, as scientists focused on identifying the building blocks of living cells and the chemical reactions that they utilize in their metabolism.

In the past century, especially in the last three decades, the rapid advances in technology and scientific discovery have tended to separate most sciences from the general public. Yet despite an ongoing trend to leave the majority of the physical sciences to the scientists, interest in human biology has actually increased among the general population. This is primarily due to medical discoveries that increase not only lifespan but also healthspan, or the number of years that people live disease free. But another important aspect of this trend is the desire among the general public to be able to ask intelligent questions of their physicians and seek additional information on prescribed medications or procedures. In many cases this information serves as a system of checks and balances on the medical profession, ensuring that the patient is kept well informed and aware of the fundamentals regarding the procedure.

This is one of the most remarkable ages in the study of human biology. The recently announced completion of the Human Genome Project is an indication of how far biology has progressed. Barely fifty years ago, scientists were first discovering the structure of DNA. They now are in possession of an entire encyclopedia of human genetic information, and although they are not yet exactly sure what the content reveals, scarcely a week goes by without a researcher announcing a medical discovery that was made possible by the availability of the complete human genetic sequence. Coupled to this are the advances in the development of pharmaceuticals and treatments that were unheard of less than a decade ago.

But these benefits to society do not come without a cost. The terms stem cells, cloning, and gene therapy no longer belong to the realm of science fiction. They represent advances in the sciences that may hold the key to increased longevity. However, in many cases they also produce ethical and moral questions of society: Where do medical researchers obtain the embryonic stem cells for their work? Who will determine if humans can be cloned? What are the risks of transgenic organisms produced by gene therapy? These are just a few of the potential conflicts that face modern society. Only a well-educated general public can intelligently survey the pros and cons of an ethical or moral decision regarding medical science. Armed with information, concerned people can participate in the democratic process of informing their elected officials of their concerns. Science education is an important aspect of citizenship, and thus the need for series such as this to present information to the general public.

This volume covers the urinary system. Although some may regard the urinary system as a minor body system, in comparison to the nervous or digestive systems, in reality it plays a crucial role in the maintenance of the internal environment of the human body. Without a properly functioning urinary system, the levels of toxic waste products in our blood quickly rise to dangerous levels, and in the process negatively influence the activity of every other organ system. In fact, many people are not aware of the complex nature of the urinary system until something goes wrong, such as kidney stones, incontinence, or cancer. In fact, the urinary system is a fascinating collection of tissues and organs, which filter approximately 180 liters of blood daily and ensure that the proper amount of fluid remains in the body. A normal, active life is simply not possible without a urinary system.

The ten volumes of the *Human Body Systems* are written by professional authors who specialize in the presentation of complex scientific ideas to the general public. Although any book on the human body must include the terminology and jargon of the profession, the authors of this series keep it to a minimum and strive to explain the concepts clearly and concisely. The series is ideal for public libraries, as well as for secondary school and introductory college libraries. In addition, medical professionals or anyone with an interest in human biology would find this series a useful addition to their personal library.

Michael Windelspecht
Blowing Rock, North Carolina

Acknowledgments

The author would like to thank her husband and son for their support during the production of this book. She would also like to acknowledge the contributions of Michael Windelspecht, Debra Adams, Sandy Windelspecht, and Elizabeth Kincaid, without whom this book would not have been possible.

Introduction

This reference volume has been designed to provide a comprehensive overview of the urinary system. It is geared to students, educators, patients, and anyone interested in the design and function of the human urinary tract. High school and college students will find it useful for cross-referencing concepts and terminology from introductory biology and anatomy courses. Teachers can use it as a guide to support classroom curricula. And patients and their family members will find it an easy-to-use reference for researching a doctor's diagnosis or investigating treatment options.

The first section of the book delves into the inner workings of the urinary system. Included are detailed descriptions of each organ—kidneys, ureters, bladder, urethra—and their collective role in urine production, storage, and elimination. Readers can follow the urine pathway from kidney filtration to bladder storage and finally to removal via the urethra. The section also describes in detail the process by which the kidneys filter waste products out of the blood and return necessary electrolytes and nutrients to the body, how the filtered urine travels to the bladder for storage, and how a combination of voluntary and involuntary nerves work together to release the urine from the body.

Next, readers are invited to take a look back at the discoveries and inventions that shaped the field of urology. This section follows the history of urinary study from the ancient mystics to the modern urologists, from Hippocrates' rudimentary urinalysis techniques to sophisticated laboratory diagnostic tools, and from the first crude operations to remove stones and cancers to modern high-tech surgical procedures performed in state-of-the-art medical facilities.

The final section covers in detail the diseases and conditions affecting the kidneys, bladder, and prostate gland. Each disease is accompanied by a description of risk factors, symptoms, diagnostic methodology, and treatment options. This section also highlights the latest developments and breakthroughs in urinary system research.

This work is designed to reach a general audience, and as such, every effort has been made to use simple, descriptive language. The reading level is appropriate for high school and introductory college students. When sophisticated terminology is necessary to describe a scientific or technical concept, it is highlighted in **bold** and a definition is provided in the glossary at the end of the book. Similarly, all acronyms are spelled out in an Acronyms chapter. For readers seeking additional information on a particular condition, a chapter ("Organizations and Web Sites") containing the addresses, phone numbers, and Web sites of related organizations has also been provided. The book finishes with an index.

INTERESTING FACTS

- About 25 percent of the total volume of blood in the body is pumped through the kidneys every minute.
- A normal, healthy adult kidney is about the size of a human fist and weighs about 5 ounces. But when cysts develop, as in polycystic kidney disease, the kidneys can grow to the size of a football and weigh upwards of 38 pounds.
- At birth, each kidney weighs about half an ounce. The kidneys don't reach their final weight until adolescence.
- The male urethra measures about 8 inches, compared with the female urethra, which may only be 1.5 inches long.
- A healthy adult bladder can comfortably hold 14–20 ounces (400–600 milliliters) of urine.
- The average adult ingests about 2.5 quarts of fluid per day and urinates between 1.5 and 2 quarts of urine each day.
- The average person urinates about 4–6 times per day. People who have severe cases of the chronic disorder interstitial cystitis may need to urinate up to sixty times a day.
- The average prostate weighs about an ounce and is about the same size and shape as a walnut. But in men who develop benign prostatic hyperplasia (BPH) as they age, the prostate can swell to the size of an orange.
- Each ureter stretches about 12 inches, and at its widest portion measures about 0.5 inches around.

Structure and Function of the Urinary System

The human body is a sophisticated piece of machinery; its organs, nerves, muscles, and tissues are designed to perform one or more of the functions necessary to keep us alive and healthy. Within the body are several complex systems, each one specialized—each one crucial in its own way. The respiratory system supplies oxygen to the blood; the circulatory system transports that oxygenated blood throughout the body; the lymphatic system protects against infection; the endocrine system produces the hormones that direct bodily functions; the nervous system gathers, stores, and transmits sensory information; the skeletal and muscular systems maintain an upright posture and mobility; the reproductive system enables procreation; and the digestive system transforms food into energy.

Once the digestive system has completed its job and the body has taken the nutrients it needs, waste products are left behind in the blood. If allowed to build up, these poisonous wastes would eventually destroy cells, tissues, and organs, causing the body to simply shut down. Fortunately, humans have a built-in mechanism, called the urinary system, to rid the body of wastes. The urinary system turns toxic materials into urine, stores and carries that urine, and removes it safely from the body.

Waste removal is the urinary system's primary responsibility, but it has other important functions as well. For example, the system must maintain the proper balance of water and chemicals—ensuring that the body is hydrated but not drowning in fluid. It does this, in part, by controlling the amounts of **electrolytes**—inorganic compounds such as sodium, potassium, magnesium, and calcium—which conduct electric currents and regulate the flow of water molecules across cell membranes. Finally, the urinary system continuously monitors and regulates the acidity of body fluids.

PARTS OF THE URINARY SYSTEM

To perform all of these sophisticated functions requires a collaboration of organs, tubes, muscles, and nerves. The primary components of the urinary system are the kidneys, ureters, urinary bladder, urethra, and urinary sphincter muscles.

The Kidneys

The two bean-shaped **kidneys** are the functional core of the urinary system (see Figures 1.1A and 1.1B). They keep the body free from impurities, maintain a healthy water and chemical balance, oversee the composition of electrolytes, regulate blood pressure, and secrete several important hormones.

The kidneys are located on either side of the spine toward the back, just underneath the ribcage. The right kidney is slightly lower than the left to make room for the liver. In an average adult, each kidney measures about 5 inches long, 3 inches wide, and 1 inch thick, and weighs about 5 ounces. Three layers of tissue encase and protect each kidney: The **renal** (renal is another word for kidney) **capsule**, a smooth fibrous membrane, forms the innermost layer. It is surrounded by the **adipose capsule**, a layer of fatty tissue. Finally, the outermost layer, the **renal fascia**, is composed of connective tissue that holds the kidney to the abdominal wall.

The outer portion of the kidney is called the **cortex** (see Figure 1.2). In the center of the kidney is the **medulla**, which contains 10–15 cone-shaped collecting ducts called **renal pyramids**. The renal pyramids drain urine into cup-shaped receptacles called **minor calyces**. From here, the urine flows into larger openings called **major calyces**, through the funnel-shaped **renal pelvis**, and on to the ureter and bladder.

Most kidney function actually takes place in microscopic cup-shaped capsules called **nephrons** (see Figure 1.3). Each kidney contains about 1 million nephrons, and it is here that the blood is filtered. Toxic wastes are removed while water and necessary nutrients are reabsorbed into the system. Nephrons also control the blood **pH** level—in other words, they make sure that the blood is neither too acidic nor too alkaline.

Inside each nephron is the **glomerulus**, a network of tiny blood vessels or capillaries, which are contained in a thin sac called the **glomerular capsule** (also called the **Bowman's capsule**). About 25 percent of the total blood pumped by the heart passes through the kidneys every minute. The blood enters each kidney via the renal artery in the **hilius** (the curved notch near

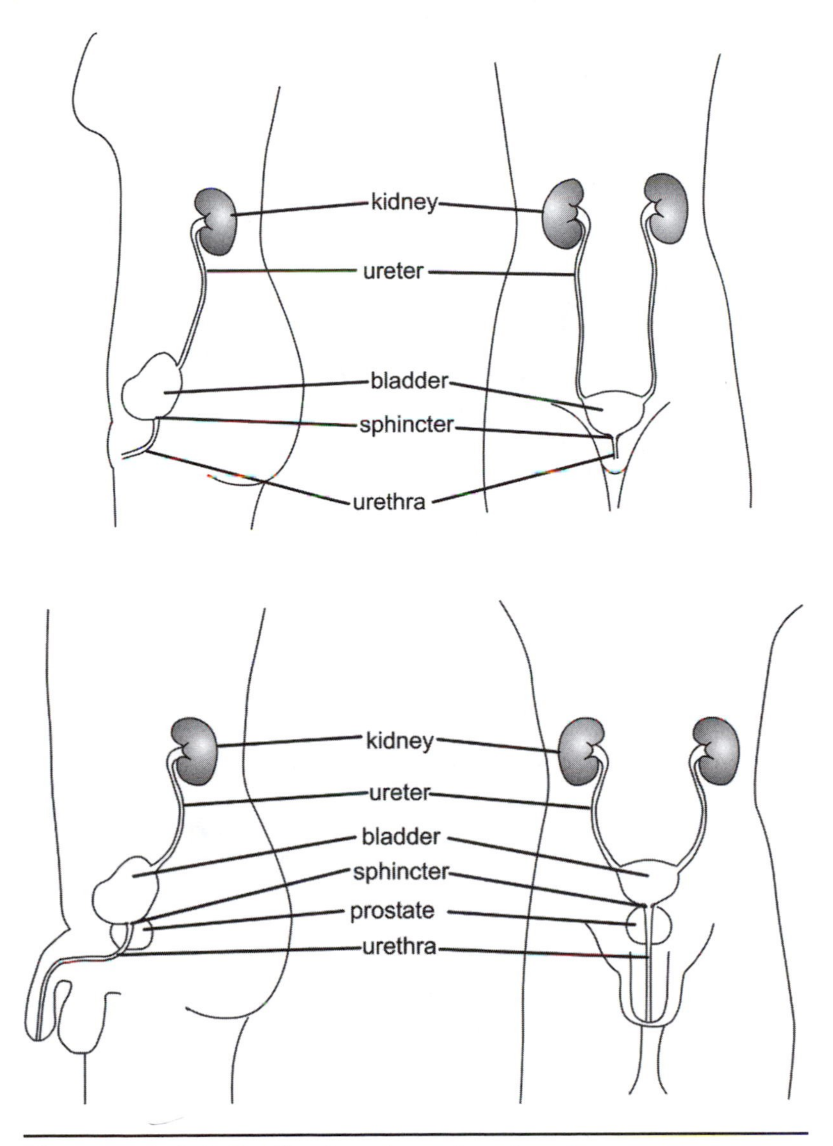

Figures 1.1 A & B. The female and male urinary systems.
The two ureters lead from the kidneys to the bladder. The female urethra is shorter (1½ inches) than the male urethra (8 inches). In the male, the walnut-shaped prostate encircles the urethra.

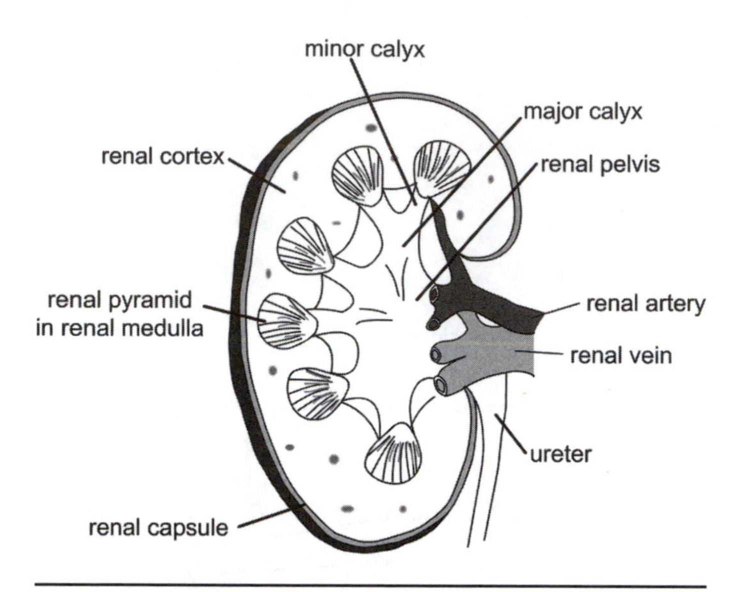

Figure 1.2. The kidney.
Filtered urine drains from the renal pyramids into the minor calyces, through the major calyces, and on into the renal pelvis. From there, it flows down through the ureter to the bladder.

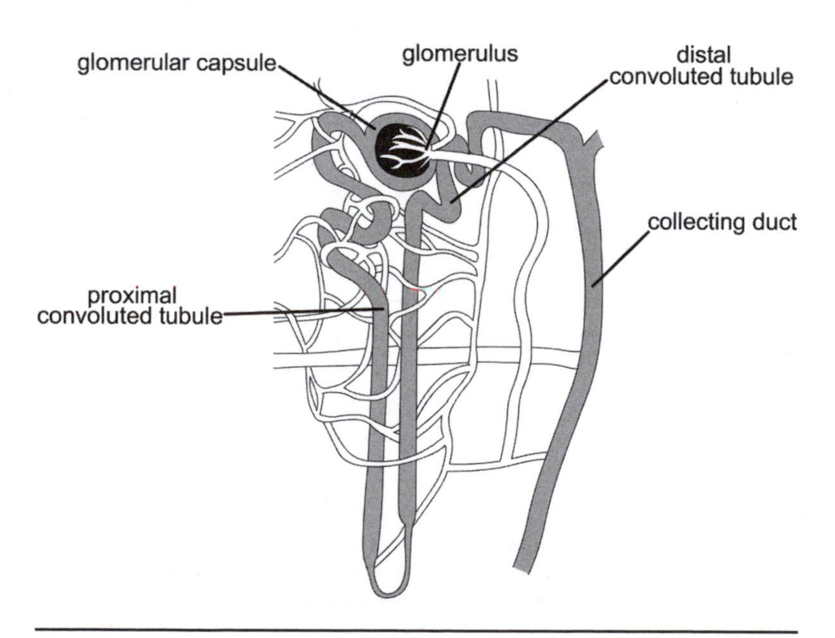

Figure 1.3. The nephron.
The blood is filtered inside the tiny capillaries of the glomeruli. After the filtrate passes through the proximal convoluted tubule and distal convoluted tubule, the waste product, called urine, is emptied into the collecting duct.

the center on the side of each kidney), then branches off into the capillaries of the glomerulus. As the blood flows out of the glomerulus, it passes through a three-layered membrane, beginning the filtration process. The membrane stops blood cells and large protein molecules from passing through, but allows water, electrolytes, sugars, and amino acids to continue into the glomerular capsule.

When the filtered fluid, called **glomerular filtrate**, reaches the glomerular capsule, it moves through a small coiled tube—the **proximal convoluted tubule (PCT)**. It is in this tubule that the real recycling work begins. Water, sodium, sugar, calcium, proteins, and other substances the body needs are reabsorbed into the bloodstream through tiny capillaries. The remainder of the fluid, containing substances not reabsorbed by the blood—such as water, **urea** (a waste product created by the breakdown of proteins), and excess salts—is carried through the U-shaped **loop of Henle**, to the **distal convoluted tubule (DCT)**. Several distal tubules empty the waste, called urine, into a single **collecting duct**. The collecting ducts in turn empty urine into larger **papillary ducts**. As it moves down the path from proximal convoluted tubule to papillary duct, the solution is continually being filtered, so that by the time it reaches the ureters, about 99 percent of the original glomerular filtrate has been reabsorbed into the bloodstream. The cleansed blood makes its way back to the heart through the renal vein, while the urine flows into the calyces of the renal pelvis on its way to the ureters.

KIDNEY FUNCTIONS

The kidneys have three main functions: homeostasis, waste removal, and hormone secretion.

Homeostasis. The primary function of the kidneys is **homeostasis**—maintaining a balance of fluids within the body. The body contains more than 40 quarts (37 liters) of fluid, which is found in and around the cells. About two-thirds is intracellular fluid, located within the cells themselves. About 75 percent of the remaining extracellular fluid is found in the tissue outside of the cells (called interstitial fluid), and the other 25 percent is contained in plasma, the fluid portion of blood.

Water passes in and out of these three fluid areas via a process known as **osmosis**. Surrounding each cell is a **semipermeable** (or **selectively permeable**) **membrane**, which separates fluids of different concentrations. The semipermeable membrane allows certain molecules to pass through while restricting the movement of other molecules. In osmosis, water moves across this membrane via a passive process called **diffusion**, from an area of higher concentration to an area of lower concentration until the two volumes are equal. The process is called passive because fluid is not pushed across the membrane by any outside force, but simply flows from higher to lower concentration.

The body takes in about 2.5 quarts (2,500 millimeters) of water every day through food and beverages. What goes in must equal what goes out, and the body has several routes by which fluid can exit the body: the kidneys (urine), skin (perspiration), lungs (breath), and intestines (feces). When a malfunction in the water removal process occurs, the body becomes overly saturated or parched. Too much water in the blood can force the heart to work harder and dilute essential chemicals in the system. Too little water, called dehydration, can lead to low blood pressure, shock—and even death. The kidneys help to balance the fluid in the body by reabsorbing liquid into the bloodstream when levels get too low, or by eliminating excess fluids when levels rise too high. These processes occur under the direction of the **hypothalamus**, the part of the brain that also regulates metabolism, body temperature, blood pressure, and hormone secretion.

If the concentration of water drops too low (because not enough liquid was ingested or because fluid was lost through sweating, vomiting, or diarrhea), neurons called **osmoreceptors** send a message to the hypothalamus, which in turn tells the pituitary gland to secrete antidiuretic hormone (ADH; also known as vasopressin) into the bloodstream. This hormone increases the permeability of the distal convoluted tubules and the collecting ducts in the nephrons of the kidneys, thus returning more fluid to the bloodstream. When more water is reabsorbed, the urine becomes more highly concentrated and is excreted in smaller volume. When the fluid concentration in the body is too high, ADH is not released. The distal convoluted tubules and collecting ducts are less permeable to water, and the kidneys filter out excess fluid, producing a larger volume of more dilute urine.

The kidneys must also maintain a balance of sodium, potassium, and other electrolytes in body fluids. To do this, they separate ions from the blood during filtration, returning what is needed to the bloodstream and sending any excess to the urine for excretion. Electrolyte levels are directed by the endocrine system, a collection of hormone-releasing glands. Hormones are chemical signals that travel through the bloodstream, triggering cells to complete a particular job.

Sodium and potassium are two of the most important electrolytes, because without them, fluids would not be able to properly move between the intracellular and extracellular spaces. Sodium is the most abundant electrolyte in the extracellular fluid, and it also plays an important role in nerve and muscle function. The presence of too much sodium (a condition called **hypernatremia**) will cause water from inside the cells to cross over into the extracellular region to restore balance, causing the cells to shrink. If nerve cells are affected, the result can be seizures, and in rare cases, coma. Too little sodium (called **hyponatremia**)—lost from excessive diarrhea, vomiting, or sweating—can send water into the cells, causing them to swell. This can lead to weakness, abdominal cramps, nausea, vomiting, or diarrhea. The swelling is even more dangerous if it occurs in the brain, where it can lead to disorientation, convulsions, or coma.

Potassium assists in protein synthesis and is crucial for nerve and muscle function. Too little potassium can lead to a buildup of toxic substances in the cells that would normally pass into the extracellular fluid. To prevent a sodium-potassium imbalance, the cells use a mechanism called the **sodium/potassium pump**. This pump is a form of active transport (as opposed to the passive transport used in osmosis), which means that fluid can pass from one side of a semipermeable membrane to another, even if the concentration is already high on that side. But active transport requires energy to push molecules across the membrane. That energy is derived from **adenosine triphosphate (ATP)**, a byproduct of cellular respiration. Once activated by ATP, the sodium/potassium pump pushes potassium ions into the cell while pumping sodium ions out of the cell until a balance is reached.

Endocrine hormones regulate the amount of sodium and potassium in the bloodstream. In the case of a sodium imbalance, an enzyme secreted by the kidneys, called **renin**, stimulates the production of the hormone **aldosterone** by the adrenal glands located just above the kidneys. Aldosterone forces the distal convoluted tubules and collecting ducts in the nephrons to reabsorb more sodium into the blood. It also maintains potassium homeostasis by stimulating the secretion of potassium by the distal convoluted tubule and collecting ducts when levels in the bloodstream get too high. Aldosterone also indirectly regulates the balance of chloride. As sodium is reabsorbed, chloride is present and is passively reabsorbed into the bloodstream.

Parathyroid hormone (PTH), produced by the four parathyroid glands in the neck, regulates levels of bone-building calcium and phosphate. When calcium concentrations in the body drop, PTH pulls calcium from the bones, triggers the renal tubules to release more calcium into the bloodstream, and increases the absorption of dietary calcium from the small intestine. When too much calcium circulates in the blood, the thyroid gland stimulates the production of another hormone, **calcitonin**, which causes bone cells to pull more calcium from the blood and increases calcium excretion by the kidneys. PTH decreases phosphate levels in the blood by inhibiting reabsorption in the kidney tubules, and calcitonin stimulates the bones to absorb more phosphate.

In addition to fluid and electrolyte balance, the kidneys play a crucial role in regulating the acidity, or pH, of fluids in the body. Water in the body is composed of hydrogen and oxygen molecules, which are held together by a chemical bond. Often the hydrogen and oxygen molecules separate into the positively charged H^+ ions and the negatively charged OH^- ions. An excess of H^+ will make the solution acidic, and too much OH^- produces an alkaline solution. Acidity is generally measured on a scale of 0–14. A neutral solution measures 7, right at the center of the scale. The higher the pH, the more alkaline the solution; the lower the pH, the more acidic the solution. The body pH must remain within a very narrow range, between 7.35 and 7.45, in order to survive. Fortunately, when the body fluids become too acidic or too alkaline, the kidneys either eliminate or reabsorb hydrogen ions until the pH returns to within its normal range.

Waste Removal. As food moves through the stomach and intestines, digestive enzymes break the nutrients into smaller particles to be used by the body. This breakdown process releases several toxic waste products into the bloodstream. These include:

Urea. Amino acids, derived from protein metabolism, are broken down in the liver to form ammonia. Because ammonia is too poisonous for the body to process, the liver converts it into the less toxic urea for removal.

Uric acid. Formed by the breakdown of purines (components of foods) in the tissues.

Ketone bodies. Produced by the breakdown of excess fatty acids in the liver.

Creatinine. A byproduct of muscle metabolism.

If any of these wastes were allowed to build up in the blood, they would eventually poison the blood and cells. The kidneys filter out dissolved wastes from the bloodstream to form urine, which is eventually removed from the body. More on urine formation later in this chapter.

Hormone secretion. The kidneys either secrete or activate three essential hormones:

Erythropoietin. Stimulates the production of red blood cells in bone marrow.

Calcitrol. Promotes bone growth by increasing the levels of calcium and phosphorous in the blood.

Aldosterone. Regulates blood pressure and sodium balance by increasing the filtration of blood in the kidneys, increasing water reabsorption, and decreasing the amount of sodium that is lost. The kidneys don't actually produce aldosterone, but they do control its production by secreting renin, an enzyme that converts a protein in the blood called angiotensin to angiotensin I. As it passes through the lungs, angiotensin I is converted into angiotensin II. Angiotensin II stimulates the release of the hormone aldosterone from the adrenal cortex.

The Ureters

Out of each kidney extends a **ureter**, a thin, hollow tube that reaches down into the bladder. Each ureter (see Figure 1.4) stretches about 12 inches, and at its widest portion measures about 0.5 inches around. The ureters pierce the bladder walls from either side, forming a U shape. At the

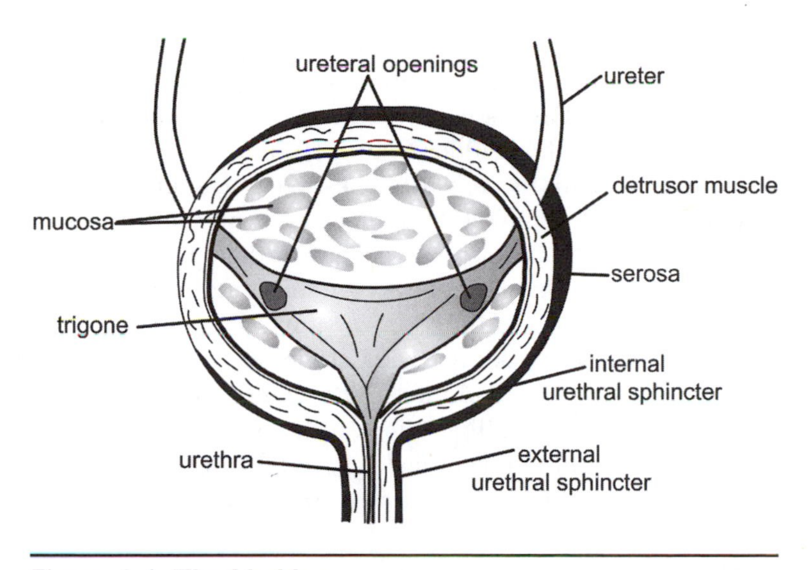

Figure 1.4. The bladder.
The ureters pierce the bladder on either side, reaching down into the organ to form a triangular shape, called the trigone. As urine funnels down the ureters and into the bladder, the detrusor muscle in the bladder wall is relaxed. When the bladder empties, the detrusor muscle contracts and the internal and external urethral sphincters relax.

bottom of the U, the ureters connect to the triangular-shaped area on the bladder floor called the **trigone**.

As the kidneys turn waste into urine, muscles lining the ureter walls help to push the urine down into the bladder for storage. Urine enters the bladder through openings called **ureteral orifices**. Valve-like mucous membranes inside the ureters keep the urine inside the bladder and prevent it from traveling back up toward the kidneys where it could cause an infection.

The Bladder

The hollow, muscular **bladder** (see Figure 1.4) stores urine until it is time for elimination. The bladder is located in the abdomen, just behind the pubic bone. As mentioned in the previous section, the ureters pierce the top of the bladder at a diagonal angle, forming a U shape where they intersect with the trigone. At the bottom of the trigone, in the neck of the bladder, is the opening to the urethra through which urine exits the body.

The inside of the bladder is composed of three layers: the **mucosa, detrusor muscle**, and **serosa**. The serosa, or outer coat, is made up of fibrous tissue. The detrusor muscle is actually a collective term for three layers of smooth muscle. This muscle is involuntary, meaning that it is not consciously controlled but is under the direction of the autonomic nervous system (see the section on urine removal later in this chapter for a description of the autonomic nervous system's role in bladder contractions during urination; for more information on the autonomic nervous system, see the Nervous System and Sense Organs volume in this series). Finally, the mucosa, or inner lining, protects the bladder from infection.

As the bladder fills with urine, it stretches like a balloon. A healthy adult bladder can comfortably hold 14–20 ounces (400–600 milliliters) of urine. When the bladder fills up, a message is sent to the spinal cord indicating the need to urinate. The bladder muscles are relaxed during filling, but they contract during urination to push the urine down the urethra and out of the body. As the bladder contracts, its walls compress, preventing urine from backing up into the ureters and kidneys. The angle of the bladder also shifts during urination. While it is filling, the bladder is tilted at the point where it attaches to the urethra to prevent leakage. But during urination, the angle shifts to allow urine to flow down the urethra and out of the body.

The Prostate

Only men have a **prostate**, the doughnut-shaped gland (see Figure 1.5) that surrounds the urethra at the point where it connects to the bladder. In young men, the prostate is about the size and shape of a walnut, but with increasing age it can enlarge to about the size of an orange.

The primary function of the prostate is to add nutrients and liquid volume to the sperm. Thousands of tiny glands inside the prostate produce a fluid that is mixed with sperm in the urethra during orgasm. The combined fluids, called semen, are released through the penis during ejaculation (the reproductive functions of the prostate will be discussed in greater depth in the reproductive system volume of this series). The prostate also protects the bladder from infection. Its muscular fibers squeeze to help control the flow of urine into the urethra.

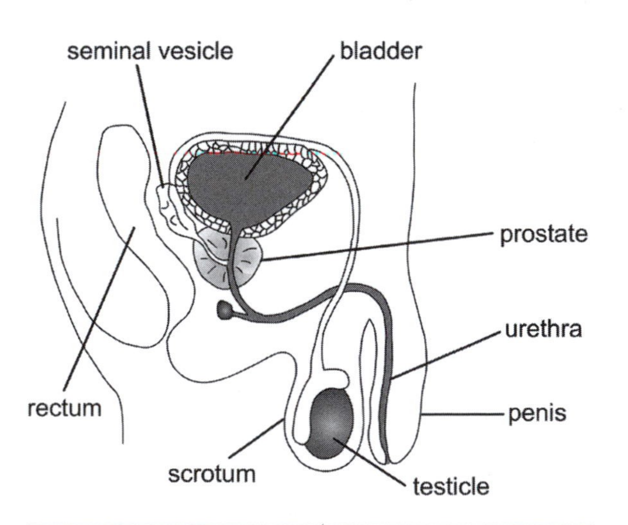

Figure 1.5. The prostate gland.
Close-up of prostate gland from point at which it connects to bladder.

The Urethra

The thin, muscular tube of the **urethra** (see Figure 1.4) connects the bladder to the exterior and provides a passageway for urine to leave the body. The wall of the urethra is made up of a mucous membrane, as well as a layer of smooth muscle tissue. The length, path, and function of the urethra differ in men and women.

In men, the urethra is about 8 inches

long and is connected from the bladder to the tip of the penis. It not only carries urine but also serves as the duct through which semen is released during ejaculation. In women, the urethra only measures about 1.5 inches in length, which is why women are much more likely to suffer a urinary tract infection than men (see Chapter 9 on urinary tract infections). This short length makes it easier for urine to slip back up into the bladder and cause infection. The urethra connects to the top of a woman's vagina, just beneath the clitoris, and its sole function is to carry urine out of the body.

The Urinary Sphincter

Two groups of muscles, called **sphincters** (see Figure 1.4), control the flow of urine out of the bladder. Where the bladder and urethra meet is a ring of smooth muscle called the **internal urethral sphincter.** These involuntary muscles stop urine from flowing back up the urethra to the bladder. At the end of the urethra is the **external urethral sphincter,** voluntary muscles that release and tighten to start and stop urine flow. When open, the internal and external sphincter muscles allow urine to exit the body. When closed, they prevent the urine from escaping.

The sphincter muscles are designed to work in conjunction with the bladder. When the bladder relaxes to allow urine to enter, the sphincter muscles remain closed to prevent leakage. As the bladder contracts during urination, the sphincter muscles relax, allowing urine to flow out of the body. If they fail to work together, urine may leak from the bladder, leading to incontinence (see Chapter 8).

WHAT IS URINE AND HOW IS IT FORMED?

Every day, about 190 quarts (180 liters) of blood plasma pass through the nephrons of the kidneys. Most of that liquid—about 188 quarts—is cleansed and returned to the bloodstream. The remainder is removed from the body as urine.

Urine formation involves three processes:

1. *Filtration.* Blood from the renal artery enters the nephrons of each kidney and passes through the tiny filtering units of the glomeruli (see Figure 1.3). Blood pressure inside the capillaries of each glomerulus pushes water, small molecules (such as glucose, amino acids, and waste products like urea), and electrolytes into the Bowman's capsule, while leaving the larger blood cells and proteins in the bloodstream. The filtered fluids are called the glomerular filtrate. The amount of filtrate that forms in the glomerular capsule of both kidneys every minute is called the glomerular filtration rate (GFR). In a healthy adult, the normal GFR is about 4.2 ounces (125 milliliters) per minute.

2. *Reabsorption.* When the glomerular filtrate exits the glomerular capsule, it enters the proximal convoluted tubules. Nearly all of the water, sodium, and nutrients such as glucose, potassium, and protein are reabsorbed from the filtrate into the bloodstream. About half of the urea is also reabsorbed; the rest is excreted in the urine. The movement of these substances occurs through active and passive transport. Glucose, amino acids, sodium, and potassium are carried across the cells of the tubules via active transport. Water is reabsorbed passively via osmosis.
3. *Secretion.* As the filtrate enters the loop of Henle and moves up the distal convoluted tubule, wastes from the blood are added. Secretion is essentially reabsorption in reverse—rather than substances moving from the filtrate into the blood, they move from the blood into the filtrate. Substances added to the filtrate may include excess hydrogen, potassium, nitrogenous wastes (urea, creatinine, and uric acid), and certain drugs (such as penicillin). After rising up the ascending limb of the loop of Henle, the filtrate moves through the distal convoluted tubule and into the collecting duct, where it is referred to as urine.

What Makes up Urine?

Because the human body is composed of about 50 to 70 percent water, it makes sense that the urine is primarily made up of water. In fact, about 95 percent of urine is water. The remainder is made up of dissolved wastes, including urea, creatinine, uric acid, and ketone bodies. The urine may also contain small amounts of substances the body normally uses, like sodium, potassium, and calcium. If the bloodstream contains excessive amounts of these nutrients, the kidneys will excrete the leftover portion into the urine. Other substances may end up in the urine that signal a problem within the body. The presence of protein or white blood cells during a urinalysis may indicate an infection or inflammation of the kidneys (see Chapter 4). Glucose in the urine may signal diabetes.

The volume and concentration of water in the urine are determined by how much water is reabsorbed as the filtrate passes through the end of the distal convoluted tubules and collecting ducts on the final leg of the urine production process, as controlled by the **antidiuretic hormone (ADH)** (see the section "How Is Urine Removed from the Body?" later in this chapter). If the body is dehydrated, more ADH is produced, making the cells more permeable to water. As more water is reabsorbed by the kidneys, the urine becomes more concentrated. If there is too much water in the body, less ADH will be produced, and water will not be reabsorbed in the distal convoluted tubules and collecting ducts, making the urine more dilute.

What Are the Characteristics of Urine?

Urine is typically yellow or amber in color. The yellow color is created by **urochrome**, a pigment produced from the breakdown of bile (the yellowish fluid secreted by the liver to digest fats). The color of urine may change depending on fluid levels in the body and on the types of foods ingested. A deficiency of fluid in the body forces the kidneys to absorb more water, creating a more concentrated, darker-colored urine. Too much fluid in the body results in a more dilute, lighter-colored urine. If the urine turns green or red, it may simply be because a person's diet included asparagus, beets, foods containing dyes, or certain medications. A change in color alone is usually no cause for concern, but the presence of pinkish or red blood in the urine (**hematuria**) can be a sign of disease (like kidney stones) or infection (see Chapters 4 and 9).

Normal, healthy urine is clear immediately after urination, but becomes cloudy when left standing. Likewise, urine doesn't have a strong smell as it leaves the body, but when left outside for any length of time, it develops an ammonia-like odor. This occurs because of the process in which the liver breaks down proteins. Referring back to the discussion on waste removal in the kidney function section of this chapter, the breakdown of proteins releases ammonia, which is too toxic for the body to handle. Ammonia is converted into the less toxic urea before it is sent to the kidneys for filtration. But when urea in the urine is exposed to oxygen in the air, it converts back to ammonia, which explains the strong odor. If the urine has a strong, foul smell immediately after leaving the body, it may be the result of bacteria from a urinary tract infection. In diabetics, urine will have a sweet, fruity smell because it contains excess ketone bodies.

The normal pH of urine ranges from 4.5 to 8.0, averaging around 6.0. The more acid the body retains, the more acidic the urine. Certain conditions may lead to overly acidic urine, including uncontrolled diabetes, diarrhea, dehydration, and **acidosis** (an abnormal increase in acidity when blood pH drops below 7.35). Urine tends to be more alkaline when an individual suffers from urinary tract obstruction or chronic kidney failure. A vegetarian diet can also make the urine more alkaline.

How Is Urine Removed from the Body?

The average person drinks between 1.5 and 2 quarts (1,500–2,000 milliliters) of fluid per day, and voids between 1 and 2.5 quarts (1,000–2,500 milliliters). Obviously, the more liquids that are ingested, the more urine the kidneys produce and eliminate. How much urine is produced also depends on the types of liquids ingested. For example, soda and coffee contain caffeine, which is a diuretic. A **diuretic** increases urine production and leads to more frequent urination than a nondiuretic like water.

If an individual urinates more than the volume of liquids ingested, the body becomes dehydrated. Prolonged dehydration can be caused by excessive sweating, vomiting, diarrhea, or extremely rare conditions like diabetes insipidus. If the fluid level in the body drops too low, serious illness may occur. Extreme dehydration can eventually result in death. Thankfully, the body has a built-in regulating system to prevent dehydration. When the level of fluid in the extracellular spaces drops, receptors in the hypothalamus of the brain release the antidiuretic hormone (ADH) into the bloodstream. ADH increases the reabsorbtion of water in the distal convoluted tubules and collecting ducts of the kidneys, creating a more concentrated urine. It also triggers that dry-mouthed feeling of thirst.

The average person urinates every 3–5 hours throughout the day and can sleep through the night without having to use the bathroom. Of course, what is normal varies from person to person. Some people may drink more, and therefore find it perfectly normal to urinate every 1–2 hours, while others drink less or have more accommodating bladders, and may only have to go every 5–6 hours. The volume of each urination can vary from about 4 ounces to 34 ounces, depending upon how much the kidneys are producing and how much the bladder can comfortably hold.

The process by which urine is released from the bladder, through the urethra, and out of the body is called **micturition**, or urination, and it occurs continuously. Every 10 to 15 seconds, small amounts of urine are forced down the thin tubes of the ureters and into the bladder. The bladder stores urine until an appropriate time—that is, until the person has reached a bathroom. As mentioned earlier, when the bladder fills, its muscles relax, but the muscles of the sphincters that surround the bladder opening tighten to prevent leakage. The process leading up to urination is controlled by the central nervous system and involves a combination of involuntary and voluntary nerve impulses. The bladder is controlled by the autonomic nervous system, which oversees other involuntary actions, like breathing. The voluntary muscles of the urethra are controlled by the somatic nervous system. As the amount of urine rises and the bladder stretches, receptors in its walls send a message to the spinal cord via sensory neurons that it is getting full. The spinal cord passes the message along to the brain in the form of impulses, and the result is that feeling of pressure in the lower abdomen that signals the need to urinate.

When it is time to urinate, **parasympathetic nerves** release the neurotransmitter acetylcholine, which causes the detrusor muscle in the wall of the bladder to contract and the internal urethral sphincter to relax. At the same time, the brain tells the sphincter muscles, which until now have been tightly holding the bladder and urethra shut, to relax. As the muscles relax, the urethra opens, and urine is allowed to exit the body.

When the Urination Process Fails

Losing voluntary control over the micturition, or urination, process is called **incontinence**. Babies naturally wet themselves because their brains and spinal cords have not matured enough to control the sphincter muscle. But if incontinence occurs in adults, it is usually the result of disease or injury to the nerves controlling the bladder, damage to the external sphincter, or infection.

The opposite problem, retention, is the inability to urinate. This may be caused by a blockage in the urethra or bladder neck, an uncontrolled contraction in the urethra, or the lack of an urge to urinate. Incontinence and retention will be discussed further in Chapter 8.

From Birth to Old Age

THE DEVELOPING FETUS

For most of the nine months a baby spends growing inside its mother, the placenta takes on the role of its developing organs. One of its functions is to eliminate wastes until the baby's own kidneys have matured. Development of the fetal kidneys and other organs begins within the first few weeks after conception (known as the embryonic stage). Initially, three layers of cells emerge: the endoderm, the ectoderm, and the mesoderm. From these three layers, every organ in the body will develop. It is from the endoderm and mesoderm layers that the organs of the urinary system arise (see photo in color insert).

Kidney Development

Within the first four weeks of life, the roots of what will become the kidneys emerge in the pelvis of the growing fetus. The kidneys develop in three distinct stages called pronephros, mesonephros, and metanephros. At their most basic stage, pronephros, the kidneys more closely resemble the organs of primitive vertebrates. They originate as six to ten pairs of tubules, which open into primary ducts. The pronephros is nonfunctioning and disappears completely by the fourth week of embryonic development, but the duct system remains to form the basis of what is to become the mesonephric system.

The mesonephros is slightly more advanced than its predecessor, closely resembling the permanent kidneys of fish and amphibians. Between the fourth and eighth weeks of development, this system will serve as the main unit of excretion for the embryo. Like the pronephros, the mesonephros eventually breaks down, but parts of its duct system reemerge to make up part of

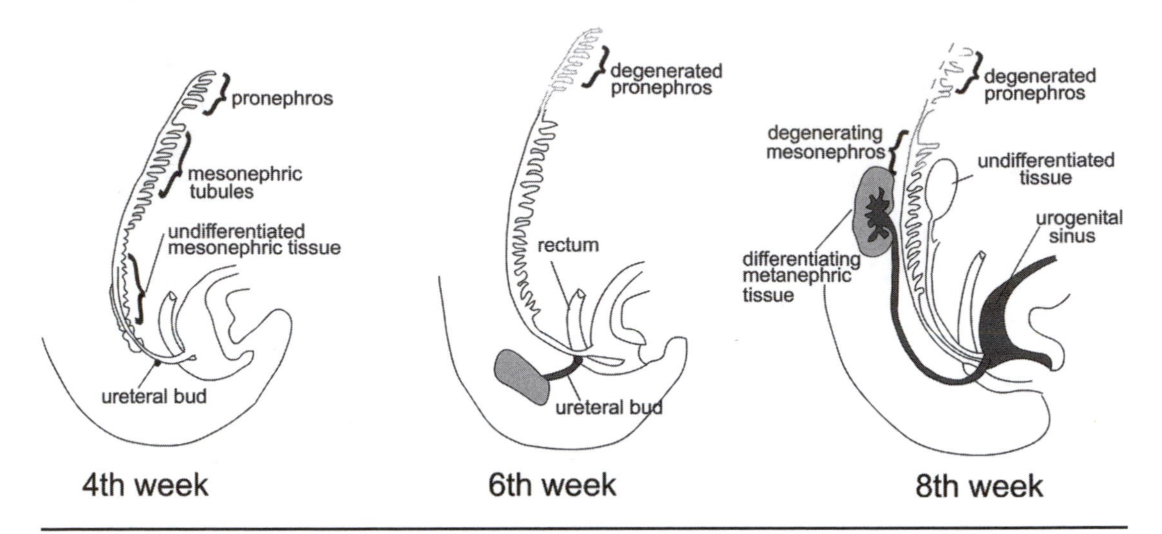

Figure 2.1. The three stages of kidney development: pronephros, mesonephros, and metanephros.
Pronephros is the first and most basic stage of kidney development. The mesonephric system, which forms the basis of the nephron, emerges out of the degenerating pronephros. During the final stage of development, called metanephros, the renal pelvis, major calyces, minor calyces, and collecting ducts emerge.

the male reproductive system. The mesonephric tubules (see Figure 2.1) develop a cuplike outgrowth, called the Bowman's capsule, into which is pushed a cluster of tiny capillaries, called the glomerulus (for more on the glomerulus, see the description of kidney function in Chapter 1). This structure will form the basis of the functioning core of the kidney—the nephron.

Between the fifth and ninth weeks of life, the final phase of kidney development, metanephros, begins. Out of this phase, the renal pelvis, major calyces, minor calyces, and collecting ducts form. As the kidneys grow, tubules develop that create the basis for the distal and proximal convoluted tubules and the loop of Henle in the glomerulus (see Chapter 1 for a description of each of these units of the kidney). By the end of the third month in utero, the tiny kidneys begin functioning, sending urine into the amniotic fluid (see photo in color insert).

Between the seventh and ninth weeks, the growing kidneys begin to slide upward toward the flank area, where they will ultimately be housed. They also rotate 90 degrees to fit their designated space. If the kidneys fail to ascend, or climb too far, the result is a condition called ectopic kidneys (see Chapter 4). An ectopic kidney may reach as high as the chest cavity, or it may cross the body to wind up on the opposite side from where it was intended. Another developmental abnormality, called a horseshoe kidney, may occur if the kidneys fuse during development and form a "U" or horseshoe shape.

Bladder Development

In the first few weeks of life, the urinary system and digestive tracts are joined by a common cavity, called the cloaca. By about the seventh week, folds of tissue fuse and divide the cloaca into two parts: the cylinder-shaped urogenital sinus and the rectum. The bladder and urethra develop together out of the urogenital sinus (see Figure 2.2). In rare cases, the cloacal membrane ruptures before the anterior bladder has separated from the abdominal wall, resulting in a congenital defect known as **bladder exstrophy** (see Chapter 6). Once the mesonephric duct from the forming kidney descends, urine begins to drain into the urogenital sinus.

One segment of the urogenital sinus forms the bladder, part of the urethra in men, and the entire urethra in women. The other portion makes up part of the urethra in men and a portion of the vagina in women. Both bladder

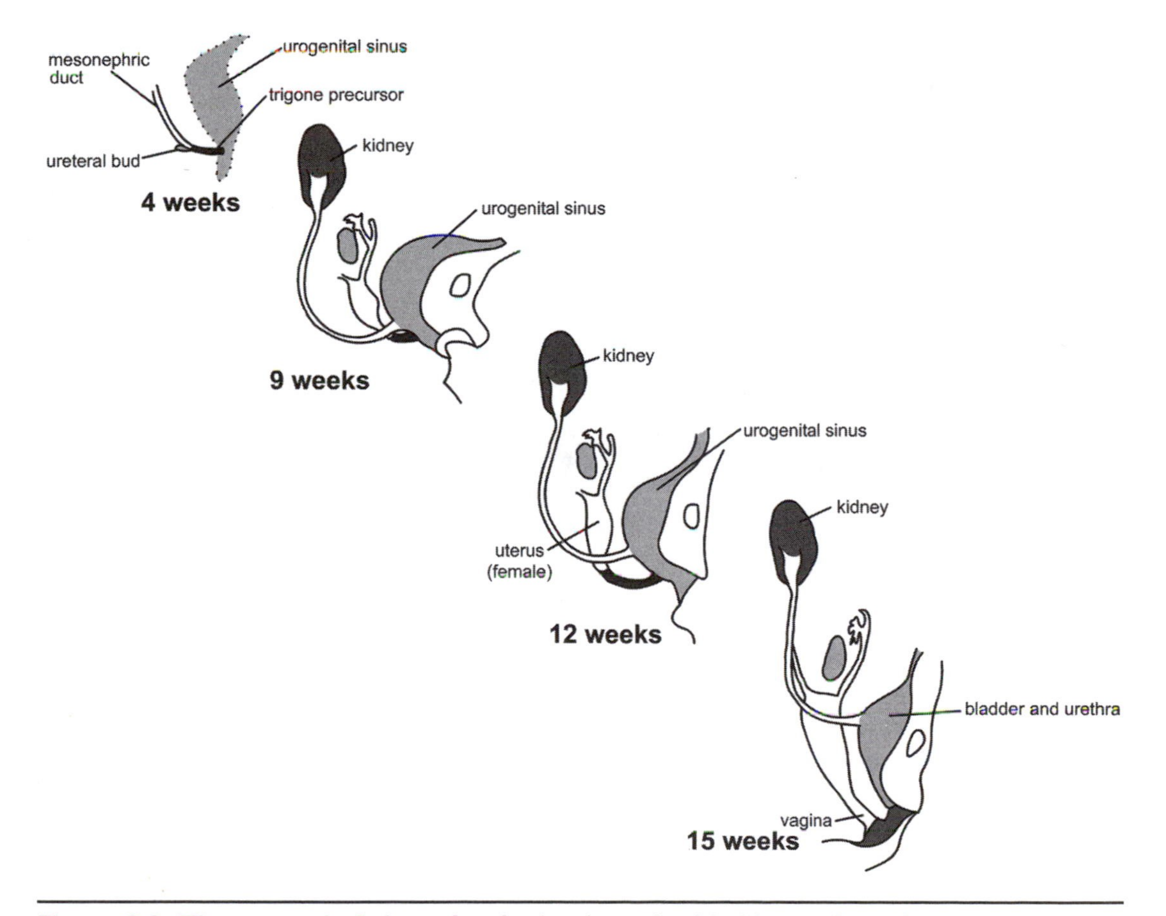

Figure 2.2. The urogenital sinus developing into the bladder and urethra.

and urethra are recognizable at around the twelfth week of development. The bladder initially extends up to the navel, where it is attached to the allantois, which connects fetus to mother. But by the eighteenth week, the bladder begins to descend and detach, and by the twentieth week, it exists as a separate entity.

Prostate Development

During the seventh week of fetal development, the male and female urinary systems are identical. But with the introduction of the hormone testosterone in the male at around the eighth week, differentiation begins, and the prostate gland begins to develop. By about the eleventh week, five groups of buds emerge out of the urethral **epithelium** (a lining of cells found throughout the body) above and below the entrance to the mesonephric duct (see Figure 2.3). Testosterone levels in the fetus rise between the thirteenth and fifteenth weeks, helping the cells mature and develop. Male hormone levels decline by the third trimester of pregnancy, and prostate growth slows.

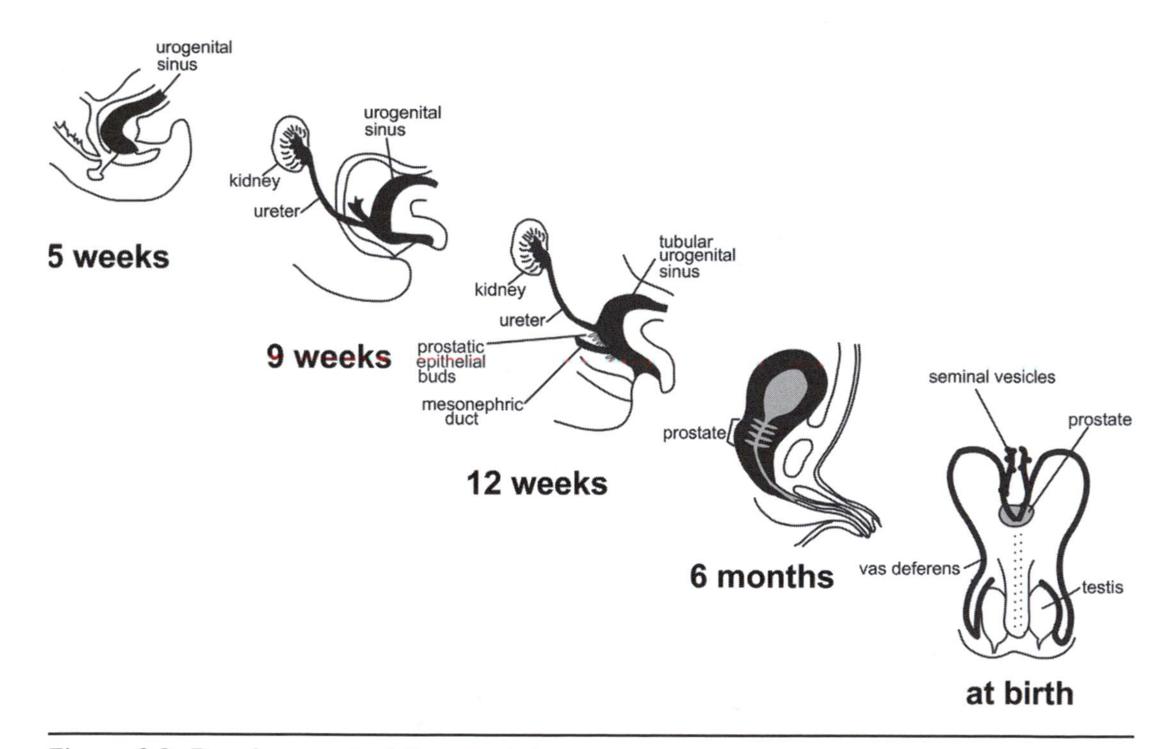

Figure 2.3. Development of the prostate.
At around the eleventh week of development, buds emerge above and below the entrance to the mesonephric duct, which develop into the male prostate.

At birth, the prostate is about the size of a pea. It remains small until the boy reaches puberty, at which time the prostate begins to enlarge. By a man's mid-20s, the prostate is fully developed and about the size of a walnut. Some men experience a period of prostate regrowth beginning in their mid-40s, resulting in a condition known as **benign prostatic hyperplasia (BPH)** (see Chapter 7 for more on conditions relating to the prostate).

BIRTH THROUGH THE TODDLER YEARS

At birth, the organs of the urinary system are tiny but fully formed. A newborn's kidneys weigh about a half an ounce each, but they are fully functional (unless the baby suffers from a congenital malformation), producing urine and sending it to the bladder for storage.

Bladder release in a newborn is an automatic process, because the brain and spinal cord are not yet mature enough to send and receive messages that the baby has to urinate (not to mention the fact that newborn babies are not mobile and cannot get to the bathroom on their own). Babies are not aware that they feel the urge to urinate, and they have no control over the process. At first, urination is a reflex. When the bladder fills with urine, the sphincter relaxes, the bladder contracts, and babies wet their diapers.

As babies grow, and their brain and central nervous system mature, they begin to understand what that feeling of pressure in their abdomen means. By the end of their second year, toddlers should have bladder control—the ability to stop the bladder from emptying at inappropriate times. Also, a toddler's bladder will have grown to the point where it can hold enough urine to avoid spontaneous wetting. Toilet training is more than a physical process, however. It is also a psychological process. First, children must realize that they are too old for diapers and understand how to translate the urge to urinate into the actual process of using the bathroom. This usually occurs by the end of the toddler years. Together, the development of brain, spinal cord, bladder, and an understanding of the urination process is enough to keep a child out of diapers, at least during the day. Nighttime control is a bit more tricky, because many young children are unable to wake up in time to make it to the bathroom. In fact, many children wet their beds for several years after they have been toilet trained during the day (more on bedwetting in Chapter 8), but this too usually ends by age 10.

THE ADULT

In a healthy adult, urination should follow the process described in Chapter 1. But occasionally, disease, injury, or congenital abnormalities can disrupt the cycle of urination. For example, pregnancy can weaken the pelvic muscles that support the bladder and urethra, leading to involuntary leakage or release of urine, called incontinence. Bladder infection or inflammation can reduce the volume that the bladder can comfortably hold. And a blockage in the urethra can slow or even halt the flow of urine. Common problems with the kidneys, bladder, and other disruptions to the urinary system will be discussed in later chapters of this book.

PREGNANCY AND THE BLADDER

During pregnancy, a woman's body undergoes a myriad of changes as it prepares to nourish and house her new baby. The abdomen and hips expand to accommodate the baby's growing bulk. The hormone balance shifts. The breasts swell and become more tender. A woman's urinary system, too, goes through many changes and adaptations to fit the needs of her growing baby. As soon as the embryo attaches to the uterus, the placenta begins producing human chorionic gonadotropin, which is released into the mother's urine. In fact, it is by testing for the presence of this hormone in urine that the pregnancy is most often confirmed.

In early pregnancy, hormonal changes in the mother's body make the need to urinate more frequent, and often more urgent. These feelings usually subside by the second trimester. But by the end of the pregnancy, frequency and urgency once again become troublesome as the weight of the expanding uterus presses down on the mother's bladder, reducing capacity just as mother and baby are producing more urine. Over a third of pregnant women suffer from stress incontinence, and many also have urge incontinence (see Chapter 8 for more on incontinence). In addition to decreasing bladder volume and increasing urine output, hormonal changes during pregnancy may also weaken the sphincter muscles, worsening leakage.

In rare cases, pregnant women may suffer from urinary retention—an urge to urinate, but the inability to produce more than a few drops. Retention can occur when the uterus angles back in early pregnancy, and the cervix presses down on and blocks the urethra. Or, it can occur in late pregnancy when the bladder is overfilled and under continuous pressure from the growing uterus. Urinary retention early in pregnancy usually resolves as the uterus rights itself, but if the condition does not improve, doctors sometimes recommend the use of a rubber, doughnut-shaped device called a **pessary** to hold the urethra open. Retention can also occur just after delivery, especially after the mother has been given an epidural (a spinal anesthetic that numbs the body from the waist down).

Leakage and retention are not the only urinary problems that plague pregnant women. Urinary tract infections (UTIs) are also a common complaint during pregnancy, affecting as many as 8 percent of soon-to-be mothers. Pregnancy-related hormones trigger a decrease in ureteral tone, slowing the flow of urine from the kidneys to the bladder, while bladder volume increases. The result is a virtual breeding ground for bacteria. Similar to UTIs occurring outside of pregnancy, the most common culprit is the Escherichia coli or E. coli bacteria. Up to 70 percent of pregnant women develop a condition called **glycosuria**, which encourages the growth of these and other bacteria in the urine. An increase in the production of the hormones estrogen and progestin also make the urine more susceptible to bacterial invasion.

If bacteria reach the kidneys, the resulting infection (**pyelonephritis**) can be painful to the mother and potentially life-threatening to her unborn baby. UTIs have been associated with intrauterine growth retardation, prematurity, and low birth weight, which is why obstetrician/gynecologists routinely test the urine of their pregnant patients for any sign of infection. Once diagnosed, UTIs are easily treatable with antibiotics like amoxicillan or cephalosporin, which are safe to take during pregnancy. Doctors tend to shy away from fluoroquinolones and tetracyclines because they can harm the fetus.

Once the baby is born, the new mother may imagine that her urinary problems are over, but this is not always the case. Pregnancy and vaginal delivery stretch and weaken the muscles of the pelvic floor that support the bladder, often resulting in a lifelong battle with urinary leakage. To keep these muscles strong and preserve continence, doctors often recommend that women exercise their pelvic floor muscles (called Kegels; see Chapter 8) throughout their pregnancy and for several months after they deliver.

THE OLDER ADULT

The normal aging process takes its toll on the human body, leaving it increasingly vulnerable to a variety of illnesses, ailments, and injuries. Like every other functioning unit of the body, the urinary tract tends to degenerate with age. After about age 40, the kidneys begin to lose some of their nephrons, which are crucial to the filtration process. The result is a decreased glomerular filtration rate, which means that the body is less able to remove wastes from the bloodstream. Loss of kidney function may also result from conditions such as atherosclerosis (a plaque buildup in the artery walls of the kidneys), diabetes, and prostate disorders, all of which are more common in old age. Any of these conditions can eventually result in kidney failure.

As they age, both men and women may feel the need to urinate more frequently, experience leakage or loss of urine (incontinence), and wake up several times each night to use the bathroom (a condition known as **nocturia**). Among seniors who suffer from decreased mobility because of a stroke, joint problems, bone fractures, or disease, incontinence can be debilitating and often humiliating. In fact, many seniors are placed in nursing homes solely due to bladder dysfunction. Others cannot understand and react to the need to urinate because their cognitive function is disrupted by Alzheimer's or other brain disorders. Other conditions, including diabetes and urinary tract infections, can affect the normal urine storage and release process. Finally, medications commonly prescribed to the elderly to treat everything from coronary artery disease to hypertension may affect the ability of the bladder to hold or release urine.

Weakening of the bladder or sphincter muscles is also common with age, especially in women who have gone through menopause. Estrogen, which is responsible for maintaining the growth and function of tissues in the urinary system, ceases to be produced during menopause. The loss of estrogen weakens the muscles of the bladder, sphincter, and urethra, leading to incontinence. Men, too, may have problems retaining continence as they age. This is primarily due to the growth of the prostate, a condition known as benign prostatic hyperplasia (BPH) (see Chapter 7). The prostate, which is normally about the size of a walnut, can grow to around the size of an orange by the time a man reaches age 70 or 80. As the prostate grows, it may block the urethra, weakening or even stopping the flow of urine.

Age also increases the risk of urinary cancers. Bladder cancer has its highest incidence among people in their 70s, kidney cancer strikes most often between the ages of 50 and 70, and about 80 percent of prostate cancer victims are age 70 or older.

3

Urology Comes of Age: A History of the Profession from Antiquity to the Twenty-first Century

THE FIRST UROLOGISTS

Thousands of years ago, our ancestors began studying the complicated inner workings of the human body. These initial explorations undoubtedly generated a barrage of questions: Why did some individuals fall prey to ailments while others remained healthy and vigorous? What caused the strange spectrum of physical symptoms that appeared with these ailments? And how could the most serious and debilitating afflictions be stopped before they consumed the body and stole the breath?

The foundations of medicine were built not upon science, but rather on the fantastic principles of magic and witchcraft. In ancient times, people believed that evil spirits or demons were to blame for all manner of sickness. And what magic wrought, only magic could cure. The sick often went to exorcists or witch doctors to drive away the demons with sacred rites and exotic potions. It was only after centuries of careful observation and experimentation that logical thinkers were able to answer the questions that had been plaguing them, and transform myth and magic into the rational science we know today as medicine.

The Egyptians

Some of the earliest representations of scientific thought are found in the writings of the ancient Egyptians. They recorded their practices and beliefs in what is believed to be the oldest surviving medical text, the Ebers Papyrus (see illustration). The papyrus, which dates back about 3,000 years before the time of Christ, describes recipes made of plant, animal, and mineral compounds for treating a variety of illnesses. And it includes some of the earliest mentions of urinary diseases, referring to incontinence, urinary retention, **cystitis**, hematuria, **prostatitis**, and **urethritis** (all of which are described in detail in subsequent chapters). The Egyptians were surprisingly advanced in their knowledge of human anatomy and disease, in part because the process of mummification required that embalmers remove human organs before preserving a body. Such removal was almost always followed by careful observation.

The Egyptians were also the first to have a medical profession and to develop a detailed process of examination and treatment. Doctors at the time specialized in the individual regions of the body. Herodotus (c. 484–425 BCE), the Greek historian, described medical specialties of the time as "some of the eye, some of the teeth, some of what pertains to the belly." Although Egyptian doctors were moving closer to what we now recognize as medicine, they still held firm to folkloric beliefs. Although they understood the nature of external wounds, they believed that illness was a form of punishment by the gods. Before treating a patient with herbs and remedies, doctors were first required to break the evil spell.

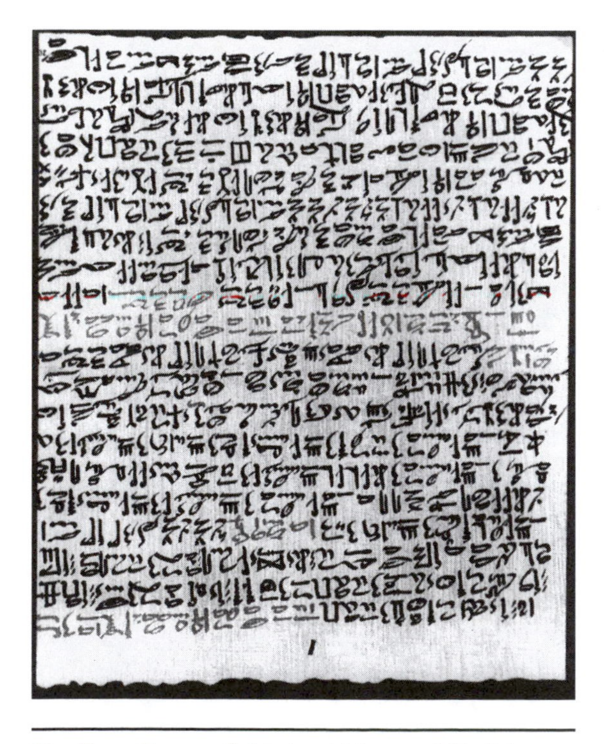

The Ebers Papyrus. © National Library of Medicine.

The Chinese

The roots of traditional Chinese medicine stretch back thousands of years, emerging virtually without influence from outside civilizations. Huang Ti (c. 2697–2597 BCE), the "Yellow Emperor," is credited with defining the Chinese system of medicine in a classic text called the *Huang Ti Nei-Ching* (*The Canon of Internal Medicine*), although the work was actually accomplished by several writers over a long period of time. The *Nei-Ching* describes the principles of harmony and balance that would form the template for both ancient and modern Chinese medical theory.

Chinese medicine was based on two essential principles—yin and yang, the balance of which was essential for good health. The Chinese believed there were five necessary organs in the human body: heart, lungs, liver, kidneys, and spleen, each of which corresponded respectively to an element: fire, metal, wood, water, and earth. The organs of the body were interrelated, like members of a family. The kidneys and bladder, for example, had the lung as their mother, the liver as their son, the spleen as their enemy, and the heart as their friend.

In addition to laying out Chinese medical theories, the *Nei-Ching* included detailed descriptions of physiological processes, including urine production. Urine was produced, according to the book, out of a mixture of fluids and a "vapor" from the kidneys. The urine was then passed through the bladder and expelled.

The Chinese were aware of urinary stones, and believed they were caused by a prolonged inflammation of the bladder, which reduced the urine to a fine gravel or sand. These particles would sometimes collect around a substance—either blood or fat—to form a mass that could weigh up to four or five pounds. After a stone formed, the patient would suffer "severe pain on voiding" and would often be "affected with retention."

Pulse-taking was the primary way in which Chinese doctors diagnosed disease, but they carried the procedure much further than today's doctors. They measured the pulse in eleven separate places on the patient's body, from scalp to ankle. The doctor would use a different finger to measure the pulse and would take the pulse measurement in a different point on the patient's body, depending upon the illness at hand. The index finger of the doctor's right hand, for example, was used to diagnose diseases relating to the kidneys and bladder. On occasion, urine was also used as an aid to diagnosis.

The principles of yin and yang factored significantly into the most popular Chinese treatment method—acupuncture. Illness was believed to disrupt or obstruct energy flowing through pathways between the surface of the body and the internal organs. Any disruption to this energy was thought to throw a patient's yin and yang out of balance. But by stimulating specific pressure points along the pathways with long, thin needles, doctors could restore balance and heal the patient. Acupuncture came into use about 5,000 years ago, and the technique remains a staple of Chinese medicine into the twenty-first century. In addition to acupuncture, Chinese medicine employed a variety of natural medicines—from white peony roots to boiled sheep's kidneys—to treat urinary afflictions as well as other diseases.

The Hindus

Like the Egyptians, the ancient Hindus wrote extensively of prescriptions, incantations, and prayers for healing a variety of illnesses, including those related to the urinary system. A scholar named Sasruta (c. fifth century) condensed the prevailing beliefs into a text known as the *Susruta Samhita*.

Sasruta's understanding of urinary system anatomy was rudimentary and still relied on mysticism, as this passage illustrates: "The tubes leaving the kidneys allow urine to flow continuously like rivers flowing into the sea; their thousand openings cannot be perceived." According to the *Susruta*, urinary diseases were caused by phlegm, bile, or air. Each of these elements left its mark on the urine: Phlegm caused it to be watery, sugary, sandy, salty, or sluggish. Bile made it acidic, or tinted it red or purple. Air turned urine the color of honey or melted butter, but this pleasant appearance belied the gravity of the patient's condition. Once the urine took on the appearance of honey, the Hindus believed, "all urinary diseases become incurable." When air and phlegm met, urinary stones were formed. These stones grew and lodged in the bladder outlet, hindering the flow of urine.

Susruta described a number of fanciful concoctions to treat urinary afflictions. For stones, he recommended a remedy made from goat's milk, camel and donkey bones, or a mixture of butter, pepper, and ginger, to which was added a few drops of sheep's urine. Urine itself was used to cure some urinary system ailments, as well as abdominal diseases and jaundice. The urine was collected from cows, sheep, elephants, as well as from humans. Hindu doctors also experimented with rudimentary surgery, using catheters to drain urine in cases of retention, and a hook-shaped apparatus to extract stones.

The Greeks

The early Greeks believed, like earlier civilizations, that illness marked the wrath of angry gods. But between 500 BCE and 500 CE, mysticism began to transform into rational science, in large part due to the efforts of the man known as the "father of medicine"—Hippocrates (460–377 BCE).

Hippocrates was the first to investigate the true nature of disease. He carefully regarded and recorded his patients' symptoms to fully understand the illness at hand. He wrote dozens of books advising other physicians how to examine their patients, document the progression of disease, and prescribe treatments. And he developed a strong code of medical ethics, called the Hippocratic Oath, with strict rules governing a doctor's conduct and responsibilities. Even today, many doctors say the oath before entering the medical profession.

Hippocrates' knowledge of human anatomy was striking for his time, especially considering the fact that dissection was not practiced. He understood that the kidneys filtered the blood and the bladder secreted urine. He recognized that blood or pus in the urine indicated a kidney disorder, and that sandy sediment was a sign of kidney stones. Hippocrates explained stone diseases as the result of drinking "waters," which deposited sand and stone in the urinary "vessels." Certain individuals were spared the agony of stones, he surmised, because their bowels were "loose and in a healthy state." He described how stones formed when the bladder was not "loose" enough to release urine. The bladder instead became heated, and the heat caused ingested substances to thicken in the urine and form a stone, which then dropped into the bladder neck and impeded the flow of urine.

To treat urinary stones, Hippocrates advocated the use of bleedings and potions made from ox dung and copper. He did not advocate surgery to remove stones, however, and in fact made the point in his oath: "I will not cut persons laboring under the stone, but will leave this to be done by men who are practitioners of this work." This condemnation may have been, in part, an expression of his concern over the dangers of such a procedure, as well as an attempt to separate legitimate physicians from the ill-reputed barber-surgeons of the day. Hippocrates did not frown on all forms of surgery, however. When a patient suffered a severe kidney infection, he advocated cutting "deeply *over* (but not into) the kidney."

Following Hippocrates' death, the center of Greek medicine shifted to Alexandria. The greatest achievement of the Alexandrian medical school was undoubtedly its contribution to the study of anatomy, led by Herophilus of Chalcedon (335–280 BCE). Now recognized as the "father of human anatomy," Herophilus dissected human and animal corpses to arrive at the first detailed description of the nervous system, duodenum (part of the small intestine), and retina. It is believed that Herophilus was also the first to distinguish between sensory and motor nerves, and he is credited with discovering and naming the prostate gland.

The Romans

The Romans, for all their military might, lagged far behind their Greek counterparts when it came to the practice of medicine. The early Romans relied almost solely on ritual and superstition, practices believed to have been borrowed from the ancient Etruscans. Treatment consisted of little more than agriculturally based home remedies, such as rams' wool to soothe uterine inflammation. Anyone who wished to practice medicine could do so without a license, and as a result, quacks ran rampant and the Roman public quickly learned to distrust their so-called medical profession.

It wasn't until the Romans began borrowing from the Greeks that they could claim a real medical profession. The first introduction to Greek medicine arrived with Archagathus of Peloponnesus, a physician who was thought to have migrated to Rome around 219 BCE. Archagathus was initially successful, but Roman traditionalists became distrustful of his surgical techniques and pharmaceutical treatments, and he was eventually expelled from Rome.

As Greek medicine was beginning to make its way to Rome, a few native Romans were also making contributions to the development of medicine at the beginning of the Christian era. The first worth mentioning is Aulus Cornelius Celsus (c. 25 BCE–50 CE). Celsus was not a doctor, but rather a writer of encyclopedias. One of his greatest accomplishments was an eight-volume medical reference called *De medicina*. His works contained descriptions of kidney stones, urologic surgery, and remedies for kidney diseases (which included warm baths, sleeping in a soft bed, and frequent enemas to release the bowels). His recommendation that patients with kidney disease avoid salty foods was remarkably astute for his time, as urologists continue to dispense the same advice to their patients today.

Pliny the Elder (23–79) was also a prolific writer. His encyclopedia of the natural sciences, *Historia Naturalis* (*Natural History*), spans thirty-seven volumes and describes in detail the fields of astronomy, zoology, botany, agriculture, and medicine. It also refers to curatives derived from plants and animals, for example the use of peony, mint, or chickpea to dissolve bladder and kidney stones. Pliny also believed that the waters of an island near Sorrento held healing powers to treat patients suffering from urinary stones.

Arguably the greatest contributor to Roman medicine was Claudius Galen (129–c. 200), philosopher, physician, and writer of numerous books and articles. Galen resurrected Hippocratic medicine for the Romans and developed theories based on Hippocrates' careful rules of observation that survived, virtually unchallenged, for more than twelve centuries. Galen was particularly interested in the function of urinary organs and the study of urine as a marker for disease. His theories on disease pathology were gained from careful observation of his patients. But to develop his advanced knowledge of anatomy, he was forced to rely on animal dissections, because Roman law prohibited experimentation on human cadavers.

Galen challenged the prevailing notion that the kidneys were functionless by asserting that they acted as a filter, purifying the blood that flowed through them from the arteries and veins. He speculated that the thicker portions of the blood were not permitted to pass through the kidneys' filters, but that the thinner portions were allowed to pass through and formed urine. From the kidney, he noted the flow of urine through the ureters and into the bladder, and he recognized that valves in the ureters prevented urine from backing up into the kidneys.

Galen delighted in questioning the prevailing inaccuracies of the day, especially those belonging to his predecessor, Asclepiades (c. 129–40 BCE), known as the "prince of physicians." As he mocked Asclepiades' mistaken belief that urine is not secreted by the kidneys, but is rather passed directly from the bladder to the bowel, Galen revealed his impressive knowledge of the urinary system:

> That these are organs for secreting the urine was the belief not only of Hippocrates, Diocles, Erasistratos, Praxagoras, and all other physicians of em-

> inence, but practically every butcher is aware of this, from the fact that he daily observes both the position of the kidneys and the duct, or ureter, which runs from each kidney into the bladder, and from this arrangement he infers their characteristic use and faculty. . . . His [Asclepiades'] view, in fact, is that the fluid which we drink, passes into the bladder by being resolved into vapors and that, when these again have been condensed, it thus regains its previous form, and turns from vapor into fluid. He simply looks upon the bladder as a sponge, or a piece of wool, and not as the perfectly compact and impervious body that it is, with two very strong coats.

Galen's writings continued to educate students and influence physicians until the Renaissance. With his death and the subsequent fall of the Roman Empire, the era of Greek and Roman predominance came to an end. Europe did not regain its medical superiority until the Renaissance.

GALENI
EXTRA ORDINEM
Classium libri, in quibus ea difiniuntur, quæ exercitatum in superioribus lectorem requirunt.
NVMERVM LIBRORVM DILIGENTIAMQVE SEQVENS PAGINA INDICAT.
Cum Decreto Summi Pont. Senatusq;
Veneti per annos. XV.
Apud hæredes Lucæantonij Iuntæ Florentini
VENETIIS M. D. XLI.

Title page of *Extra Ordinem Classium Libri,* with border illustrations showing medical scenes from the life of Galen, published 1541. Courtesy of the Library of Congress.

The Middle Ages and the Rise of Islamic Medicine

As Europe descended into the Dark Ages, the center of medicine shifted once again, this time to the Arab world. It should be noted that the world of Islamic influence, which stretched from India to Spain, encompassed not just Arabs but Europeans, Christians, and Jews as well. During this period, physicians began moving away from spiritual medicine and toward the more rational system of diagnosis and treatment originated by Hippocrates and Galen. They translated the works of these two masters and used them to form the basis of their own system of medicine.

Hippocrates and Galen were both revered, but the latter undoubtedly had the greatest influence among Arab physicians. One notable exception to this was Abu Bakr Mohammed ibn Zakariya al-Razi, better known as Rhazes (c. 865–925). Rhazes was born in Persia and practiced medicine throughout Baghdad, Cairo, and Cordova. Unlike his counterparts, he followed Hippocrates and adopted his rule of careful observation, for which he became known as the "Arabian Hippocrates." Rhazes translated and incorporated Greek as well as Roman medical theory into his own observations. Together they formed the basis of his encyclopedia, *Kitab al-Hawi fi al-tibb* (*The Comprehensive Book on Medicine*), which became one of the most popular medical texts in medieval Europe. Rhazes also wrote treatises on kidney and bladder stones and diabetes. He advocated the use of simple drugs, such as wormwood, birthwort, and scorpion oil, to treat small stones, and he recommended that doctors extract large stones via an incision at the base of a man's penis.

Another well-known Arabic physician was Ibn Sina, who is more commonly known by his Latin name, Avicenna (980–1037). Born near the town of Bukhara in Persia, Avicenna began practicing medicine at the age of 16. His goal from the very start was to develop a complete, rational system of medicine. The foundation of his system came from Greek medicine, specifically Galen, but his own work would eventually rival that of his mentor and earn him fame throughout the Islamic world and Europe. By the age of 21, Avicenna had penned the five volume *Al-Qanun fi al-tibb* (*The Canon of Medicine*), which was a staple in medical schools throughout Europe and Asia for centuries after his death. The books condense the medical knowledge of Avicenna's day, providing general discussions of the human body and its afflictions, as well as an overview of symptoms, diagnosis, and the use of herbal medicines. He described the kidneys in greater detail than Galen, but continued his predecessor's mistake of placing the left kidney lower than the right (the opposite is actually true). The books also contained descriptions of symptoms and diagnoses for bladder diseases, urinary retention, incontinence, diabetes, and kidney diseases.

Avicenna was a great proponent of urine study as an aid to diagnosis. He offered the following guide to urine analysis in his work, *Al-Arjuzat fi'l-tibb* (the *Poem on Medicine*):

> White urine witnesses the quantity of ingested food and drink; it is a sign of bad digestion. . . . Somewhat yellow, it indicates the presence of a certain quantity of bile. . . . If the sediment has decomposed blood, there is a phlegmonous tumor. . . . If one sees sand in it, be aware that there is a calculous [stone]. . . . If the urine has no odor, it is that the food has not been digested or has been digested raw. The degree of decomposition agrees with the intensity of the odor of the urine. If this odor is dreadful, be aware that the illness is in the bladder.

The Renaissance and the Advancement of Urinary System Anatomy

Leonardo da Vinci (1452–1519), the legendary painter, sculptor, and inventor, has never been known for his contributions to the field of urology, yet his understanding of the urinary system was quite remarkable for his time. Leonardo was the first modern anatomist, completing in his lifetime over 750 sketches detailing the inner workings of the human body. He drew with incredible accuracy the organs, muscles, and bones, including a finely detailed rendering of the kidney with the renal artery, vein, and ureters correctly placed. But although Leonardo's drawings were precise, he lacked a thorough understanding of urinary system physiology. The pope forbade the use of human cadavers, so Leonardo was forced to base most of his drawings on animal studies. Thus, he could draw the kidneys and bladder, but he never truly understood how they functioned.

Andreas Vesalius (1514–1564), another Renaissance-era anatomist, was the first to challenge the Galenic theories on anatomy. Although he began his anatomical studies on animal corpses, Vesalius gradually began to realize that the only way to truly understand human anatomy was to look inside human bodies. The Catholic Church remained adamantly opposed to human dissection, so Vesalius was often forced to scavenge in graveyards for suitable subjects. In 1540, after dissecting a number of bodies, he shook the establishment by daring to refute Galen's teachings. It was a bold move, and one that greatly advanced the study of anatomy. Vesalius first attacked Galen for having based his anatomical studies on the barbary ape and other animals rather than on humans. And, he challenged Galen's practice of viewing dissections rather than conducting them himself. Vesalius refuted (among other things) Galen's assertion that urine was filtered, sieve-like, in the upper region of the kidneys. He was also the first to realize that the left kidney was not lower than the right, as Galen and other anatomists had mistakenly asserted. In his famous work, *De Humani Corporis Fabrica* (On the Fabric of the Human Body; 1543), Vesalius included more than 200 woodcut illustrations showcasing the human body in greater detail

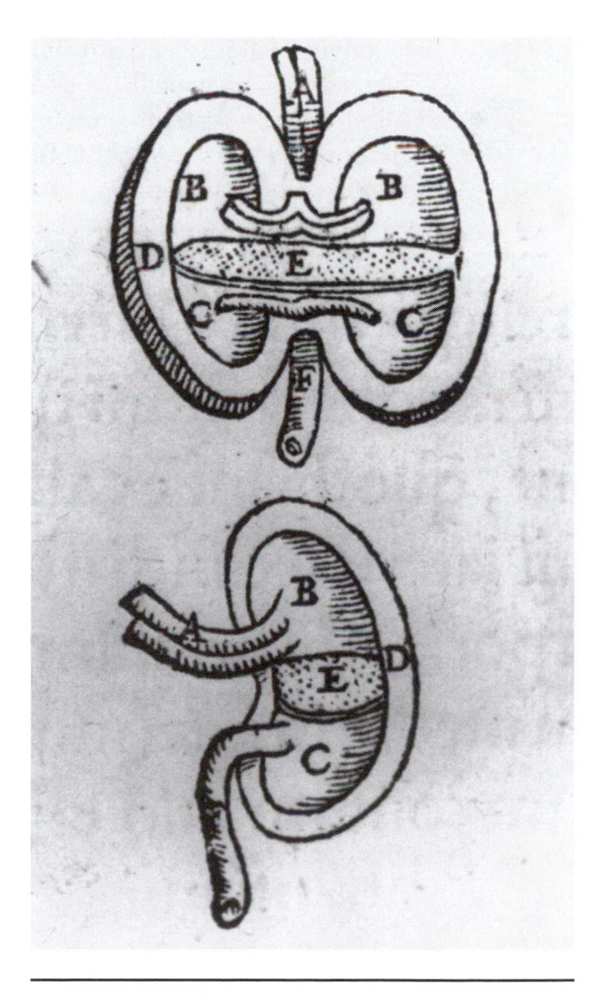

Vesalius, Andreas. Vignette of the kidney (1543). © National Library of Medicine.

than it had ever been seen before. Some consider it the most influential medical book of all time. In his short life, Vesalius managed to revolutionize the study of anatomy.

Bartolomeo Eustachio (c. 1520–1574), a professor of anatomy at the Collegio della Sapienze in Rome, did not agree with his contemporary. He wholeheartedly defended Galen and criticized Vesalius's work. Although Eustachio is better known for his work on the ear (including the tube that bears his name), he published a remarkably detailed treatise on the kidneys in his *Opuscula anatomica*. This work, which was based on his dissections of fetuses and deceased newborns, includes the most adept descriptions of those organs to date, including the discovery of the adrenal gland (suprarenal gland) and the correct assertion that the right kidney was lower than the left.

Italian anatomist and physiologist Lorenzo Bellini (1643–1704) further enhanced the understanding of kidney design and function. At age 19, Bellini published his best-known work, *Exercitatio Anatomica de structura et usu renum*, in which he asserted that the kidneys are composed of ducts, which now bear his name. Bellini also determined that blood was filtered in what we now know are the glomeruli of the kidneys. His studies were furthered by Marcello Malpighi (1628–1694), physician and professor of anatomy at the University of Pisa. Using a primitive early microscope, Malpighi was the first to discover the function of the glomeruli in secreting urine. He proved that the kidneys were more than mere filters—that they modified as well as cleansed the blood.

Malpighi made some of the earliest forays into microscopic science to describe kidney function, but filtration was not truly understood until two centuries later, when the microscope was fine tuned. The greatest progress in this area was made by a British physician, Sir William Bowman (1816–1892), who discovered the role of the glomeruli in filtering the blood and described the capsules in the nephrons that now carry his name (see Chapter 1 for a description of the Bowman's capsule).

URINE STUDY

For centuries, people have been fascinated by urine for its power to predict and even treat disease. In ancient times, urine was thought to be imbued with spiritual powers. Rubbing it on the body or adding it to a magic potion was believed to keep evil spirits at bay. Later, urine was adopted for cosmetic use, on occasion to soften the skin. As distasteful as it sounds today, the ancient Romans used urine as a mouthwash to prevent tooth decay. And the Egyptians treated diseases with various concoctions containing human and animal urine.

Eventually, doctors realized that the composition of urine was related to an individual's health. The first mention of **uroscopy**, or the examination of urine to diagnose disease, emerged in ancient Greece. Hippocrates regularly inspected and tasted his patients' urine, and he refers to urine study numerous times in his writings: "The urine is best when the sediment is white, smooth, and consistent during the whole time . . . the most deadly of all kinds of urine are the fetid, watery, black, and thick" and "Urine with a fatty upper part and a sediment indicates fever; blood-stained urine passed at the onset means a lengthy illness. Disturbed urine in association with sweating denotes a relapse; white urine like that of beasts of burden indicates headache." Hippocrates also noted the relationship between "bubbles on the surface of the urine" (a sign of **proteinuria**) and kidney disease.

Galen adopted Hippocrates' teachings on uroscopy and added a few of his own thoughts to the study. But after Galen, it was several hundred years before the practice once again came into use, revived by a physician named Theophilus Prostospatharios (610–641). Prostospatharios heated urine and examined the resulting sediment to diagnose a variety of ailments. His method of urine study borrowed heavily from Hippocrates, as is evidenced by this passage: "when the urine in consistence and color appears exactly like oil of a dark color, it prognosticates a collapse and death."

French physician Pierre Gilles de Corbeil (c. 1200) popularized uroscopy with the publication of his now classic book, *Liber de Urinus*. The work classified diseases based on changes in the urine, and proposed guidelines for urine examination. Corbeil instructed physicians to use a clear, bladder-shaped glass to hold the urine and to make their observations in just the right light—neither too bright nor too dark. Corbeil was prophetic in writing that "urine is composed of the residue left in the blood and other humors in the kidneys," as scientists did not truly understand kidney filtration until about 600 years later.

By the Middle Ages, uroscopy had become central to the doctor's repertoire. In fact, aside from pulse-taking, few other diagnostic tools existed at the time. Uroscopy became an integral part of every examination and was used to diagnose everything from illness to pregnancy. Doctors evaluated a patient's urine based on four factors: color, consistency, odor, and nonsoluble sediments or particles, color being the most important element. The matula, or transparent urine glass Corbeil had described, became a symbol of the medical profession and a status symbol among physicians.

Numerous paintings and woodcuts in the fifteenth through the seventeenth centuries (see illustration) depict the art of "water-gazing"—a physician holding a matula up to the light while the patient looks on. But in fact, the patient's presence was not necessary for a diagnosis. Sometimes, a messenger brought the patient's urine to the physician before or in lieu of an examination. Uroscopists were warned in these cases to be wary of tricksters trying to test their

Teniers, David. Diagnosis through urine examination, seventeenth century. © National Library of Medicine.

skills or deceive them by delivering the urine of another patient or even an animal. Bernard de Gordon (1285–1318), a teacher of medicine at the University of Montpellier in France, advised physicians to "look carefully at the bearer, for if he is trying to trick you, he will change colour."

While uroscopy gained favor and legitimacy among the medical profession, con artists threatened to derail the practice by hoodwinking unsuspecting patients in towns throughout Europe. These so-called water-casters, water-diviners, or doctors of urine would set up a stand in an unsuspecting village and claim that they could diagnose any illness from a patient's urine sample—for a fee, of course. The charlatans would then leave town—long before their unwitting victims could realize their folly. One sixteenth-century lawyer referred to these masters of deceit as "empirics, pests, and mountebanks," and to their victims as "stupid, conceited simpletons and complete fools."

Uroscopy flourished as an art for more than 400 years, but it was not until the 1600s that it began to transform into a science. Urine analysis progressed from simple visual inspection to chemical analysis in large part through the efforts of Belgian physician Jan Baptist van Helmont (1577–1644). Considered one of the earliest modern scientists, van Helmont was the first to investigate the chemical makeup of urine. He compared the concentration of urine samples before and after patients drank water, and looked at the relationship between urine concentration and disease. His disciple, Francis de le Boë (1614–1672), believed that chemistry was the key to good health. Illness, he thought, occurred when there was an imbalance of acid and alkali in the body. He also added another dimension to urine analysis—sound. According to de le Boë, the sound of urine passing could be distinctive or silent, depending on its consistency.

British professor Thomas Willis (1622–1675) claimed that distillation was the most precise method of urine study. He proposed that doctors separate urine into its five elements: alcohol, watery liquid mixed with sulphur and salt, spirit of urine, salt, and earth—to determine a patient's condition. And he made an unusual discovery—the urine of diabetic patients had a sweet taste. A century later, scientists would discover why.

Lorenzo Bellini's technique for urine study involved all of the senses. He smelled it, tasted it, analyzed its sound, and examined its color. Then, he evaporated rather than distilled it to separate out its individual components. Bellini believed that changes in color, odor, and taste were related to the balance of water, salt, and earth in urine.

With the discovery by Dutch physician Hermann Boerhaave (1668–1738) that urine was composed of filtered substances that previously existed in the blood, scientists turned their attention to isolating these substances. Frederick Deckers (1648–1720), a professor of medicine at the University of Leyden in Holland, used heat to separate the protein albumin from urine, and realized that its expulsion in the urine was somehow related to kidney disease (see Chapter 1). In 1798, the first chemical test to detect albumin in urine was devised. Other scientists were able to isolate phosphorous, urea, and uric acid from the urine.

In the late 1700s and early 1800s, scientists isolated the sweet taste Thomas Willis had identified in diabetic urine as sugar, devised a yeast test to detect this sugar, and revealed that the sugar was glucose. In the 1800s, scientists were also able to differentiate alkaline from acidic urine with the introduction of litmus paper—small strips of paper treated with an organic dye that turned red when submerged in an acidic solution and blue when submerged in an alkaline solution.

The first person to use **urinalysis** to diagnose and treat disease was Richard Bright (1789–1858). The British doctor and pathologist discovered the relationship between **edema**, proteinuria, and kidney disease (see Chapter 4 on kidney disease), and he confirmed that the presence of albumin in the urine indicated kidney disease. To detect urinary albumin, Bright put a bit of urine on a spoon and held it over a lit candle. "If albumin is present, you perceive before the fluid reaches the boiling point that it becomes opaque, sometimes presenting a milky appearance at the end of the spoon, which extends inwards till it meets in the centre and then breaks into a white curd," he wrote. For a number of years, kidney diseases were labeled Bright's disease in his honor.

During the 1800s, urinalysis rapidly caught on as a diagnostic tool, but it was still such a long and laborious process that it remained impractical for everyday use. Physiologist Austin Flint, Jr. (1836–1915), aimed to make things easier for doctors by standardizing urinalysis instruments and procedures in his 1870 work, *Manual of Chemical Examination of the Urine in Disease*. But what brought urinalysis into every doctor's office was the invention of the dipstick in the 1950s. These small, chemically treated strips of paper changed color based on the composition of a patient's urine. The twentieth century brought a number of other innovations, including the urea clearance test for kidney function and the "clean-catch" midstream urine test for detecting urinary tract infections (see Chapter 4 for more on urinalysis techniques). Today, doctors can quickly and easily analyze a patient's urine, breaking it up into its various constituents to determine whether the kidneys are functioning properly.

LITHOTOMY, OR STONE REMOVAL

The first lithotomists, or stone surgeons, began practicing their trade more than 4,000 years ago. Early stone sufferers were undoubtedly relieved to be delivered from their excruciatingly painful ordeal, but without benefit of anesthesia or antiseptic, the surgery itself must have been a form of torture. Consider this description of stone removal from the Hindu *Samhitas*:

> After purging the patient's body of evil spirits, the surgeon reassures the patient, and if he is brave and calm, the surgeon tells him to sit on a smooth board and to roll back with his hips raised. The patient places his knees and elbows together and these are secured with bonds. The left side of the umbilicus is then rubbed with oil and the surgeon presses firmly with his fist downwards until the stone is as low as possible. The left index and middle fingers, with cut nails, are dipped in oil and introduced into the rectum, and then are pressed forwards until the stone is grasped and stands out like a tumor.

Lithotomists at the time were clearly aware of the potentially deadly nature of this surgery:

> If, when the stone is grasped, the patient faints and his head drops like that of a dead man, the surgeon must abandon the operation, for if he proceeds, the patient will surely die.

During the operation, the surgeon was advised to proceed gingerly to protect the patient:

> Keeping a knife's breadth to the left of the raphe, the surgeon makes an incision in keeping with the size of the stone. He may make his incision on the right side, if more convenient for the operation. The surgeon must take great care not to break or even scratch the stone. A grain of dust, however small, left in the wound would begin to grow larger again.

Historians do not believe the ancient Egyptian physicians practiced lithotomy, in large part because they could find little evidence that Egyptians suffered from this affliction. Out of the thousands of mummies archaeologists have unearthed, only a handful have contained stones. One such find was a teenage boy with a uric acid stone in his pelvic bones. The composition of the stone makes sense, as the Egyptians were hunters and most likely ate a great deal of meat (see Chapter 4, for a description of uric acid stones). Another stone was found in a tomb of the Second Dynasty (c. 3000

BCE), and a third was found in the nasal cavity of a priest of Amon (c. 1000 BCE). The unusual placement of the stone in the third mummy was probably due to the embalmers, who often removed and then replaced organs in places other than where they were found.

Stone surgery was also uncommon in ancient Greece. Hippocrates prohibited the practice, fearing it far too dangerous. He was willing to make an exception only if the stone became infected and ruptured. Surgeons did remove bladder stones at the time, however. The method for stone removal used in ancient Greece and Rome is strikingly similar to that described by the Hindus, which leads to the assumption that it was introduced to the region during the conquest of Alexander the Great. One can see the similarities to the Hindu procedure in this description written by Celsus in the seventh volume of his work, *De medicina*:

> The surgeon, whose nails should be carefully pared, dips the index and middle fingers of the left hand in oil, and introduces them gently into the anus, one after the other; he presses the fingers of the right hand on the lower abdomen but only gently, to avoid injury to the bladder by pressing too hard on the stone. . . . The stone must be sought for near the neck of the bladder, and, if found, will be easy to remove. . . . When it has been found, it must be brought into position with due care if it is small and light, so that it does not slip away and one does not irritate the bladder too much. . . . Once the stone is engaged, an incision is made in the skin over the bladder neck, near the anus, down to the bladder neck, in the shape of a crescent with the ends of the wounds turned a little towards the thighs. Then in the deepest and narrowest part of this incision, a second transverse incision is made, opening the bladder neck so that a hole allowing the urine to escape is made and the opening is larger than the stone. . . . When the incision is made, the calculus is found; if it is small, it is pushed to one side by the fingers in the anus and extracted from the wound by the fingers of the other hand; if it is large, it is removed with a hook made for this purpose, applied to the upper surface of the stone.

Lithotomists followed Celsus's method, with few modifications, until the eighteenth century. Ammonius of Alexandria (b. 276 BCE) did make one improvement to the technique by breaking the stone with a blunt instrument for easier removal. It was an idea well ahead of its time, and one that would not gain full acceptance for many centuries.

During the Middle Ages, respected physicians would have little to do with stone surgery, as they believed it carried too great a risk of death. Doctors instead used medication and diet to relieve discomfort and urinary obstruction, and only resorted to the knife as a last measure. Lithotomy was left in the hands of the ill-reputed barber-surgeons of the day.

But gradually, more and more doctors were willing to attempt the practice, and accounts of stone surgeries began popping up across France, Germany, and England. The first reported account of a successful surgical stone removal was the so-called franc-archer of Bagnolet case. An archer from Bagnolet, France, was condemned to death for crimes he had committed. Upon hearing that the man suffered from chronic urinary stones, doctors asked the king and his magistrates for permission to operate. Their request was granted, and the operation was successful. The archer, cured of his affliction, was granted forgiveness for his crimes and released. He survived for many years, and the tale of this case became famous throughout Europe.

Stone surgery began catching on during the Renaissance, in part because Vesalius and other anatomists improved knowledge of kidney and bladder physiology. Lithotomists formed a licensed guild and held themselves to strict professional standards. By the end of the seventeenth century, lithotomy was considered a respectable endeavor, and its practitioners flourished.

The preferred method of stone removal at the time—the "apparatus minor," or lesser operation—was passed down virtually intact from Celsus's day. The doctor would insert his fingers into the patient's rectum and manually push the stone into the **perineum** (the area that lies between the scrotum and anus in men, and the vagina and anus in women). Once the stone was properly located, the surgeon would make an incision and use forceps to remove it. The operation was quick (a few minutes), imprecise, and difficult to accomplish on men because the prostate blocked a clear path for stone removal.

But a new method, developed by Giovanni di Romanis of Cremona, would soon supercede the "apparatus minor." To aid doctors in finding and removing the stone, Romanis introduced a guide into the bladder. The operation became known as the "grand appareil," or "apparatus major." Romanis's student, Marianus Sanctus de Barletta (1490–1550), modified the technique, and it became known as the "Marian Operation." In this technique, the patient was held by an assistant on a table with his legs bent and open, while the doctor inserted a sound into his body to verify the presence and location of the stone. Then, a grooved staff was inserted through the urethra into the bladder, widening the urethral opening. A broad knife called a novacula cut through the tissue as far as the bladder neck, and forceps were introduced to remove the stone. By the end of the sixteenth century, the "apparatus major" had become the most common lithotomy method.

But the operation was not without its setbacks. While the apparatus major was more precise than Celsus's method, it often tore the bladder, which ultimately left the patient engaged in a lifelong battle with incontinence. Another, potentially less risky design was introduced in 1556 by French surgeon Pierre Franco (1500–1570). While operating on a child to remove a stone the size of a "hen's egg," Franco could make no progress. Seeing that the child was in desperate pain, Franco abandoned the perineal approach and tried making an incision above the pubic bone, slightly to one side of the stone. He was able to remove the stone, the patient was cured, and the suprapubic method of stone removal was born. It would still be some years before the technique took hold, and even Franco was wary of his own success. He continued to use the apparatus major in subsequent operations, and advised his colleagues to do the same.

About one hundred years later, another technique was added to the lithotomist's repertoire—this one developed by an eccentric character named Frère Jacques (1651–1714). Jacques earned the title "brother" not by entering the monastery, but because he chose to wear the dark cloak of the monks. As a learned surgeon, he was skilled in both the apparatus major and minor operations. Jacques traveled through Provence, France, setting up traveling operating clinics in which he could show off his lithotomy technique. He modified existing surgical techniques by cutting laterally through the bladder. Once the bladder was opened, Jacques removed the stone with a forceps or his fingers. This approach allowed better access to the patient's bladder while avoiding the tearing of previous operations.

Jacques' reputation earned him the respect of the French king, and he was permitted to practice his craft in Paris. Unfortunately, once established in a city where he could not make a hasty exit, his mistakes began to catch up with him. During a four-month period, he operated on sixty patients and lost twenty-five soon after surgery. The French court heard of his troubles and forbade him from operating. Essentially exiled from his country, Jacques traveled through Germany and Holland, modifying and improving his operation. When he returned to France and was reinstated by the court, he had much greater success—thirty-eight lithotomies without a single death. By the end of his career, Jacques was said to have operated on 5,000 patients, some of them royalty. His technique was carried on by other surgeons, in part thanks to the work of Jacques' bitter rival, a Dutch lithotomist named Johann Rau (1658–1709). Rau helped the lateral approach catch on throughout Europe.

The next leader in the field of lithotomy was William Cheselden (1688–1752), a British anatomist and surgeon. Over the course of his career, Cheselden attempted various lithotomy techniques. First, he tried the Marian operation, or apparatus major. In 1722, he adopted Franco's suprapubic approach, only to find the tearing too grisly, especially on an unanesthetized and struggling patient. Finally, he went back to Frère Jacques' lateral method through a perineal approach, but he reversed the incision, starting it instead at the bladder and working downward to the perineum. Once the incision was made, Cheselden felt for the stone with the edge of his forceps, and, having found it, grasped it between the two blades and extracted it. To prevent excess bleeding, he would tie "the blood vessels by the help of a crooked needle, and use no other dressing than a little bit of lint besmeared with blood, that it may not stick too long in the wound." His technique was not only a success, but it also reduced the operation time from over an hour to under a minute. Crowds, fascinated by Cheselden's lightning-quick surgery, flocked to watch him work.

Lithotomy was making rapid progress, but practitioners were still reluctant to cut into the kidneys, fearing that the results of such an undertaking would be far more deadly than the natural course of the disease. This fear led many doctors to investigate other means of dissolving stones, and the result was a strange assortment of concoctions, with ingredients ranging from vinegar to lemon juice. In the 1700s, Joanna Stephens became well known for her stone-dissolving witches' brew of eggshells, snails, old clay pipes, soap, and carrot seeds. Most physicians scoffed at this remedy, but a few proclaimed it an "infallible cure."

In 1871, Gustav Simon (1824–1877) of Heidelberg performed the very first kidney removal, or **nephrectomy**. The operation was a success, although the wound took several months to heal properly. Nine years later came the first recorded **nephrolithotomy**, or stone removal surgery, on an otherwise healthy kidney. After removing a one-ounce stone from a patient, surgeon Henry Morris wrote, "This case demonstrated for the first time that a stone could be removed by cutting freely upon it through a thick layer of renal tissue." He claimed his work was the beginning of a more conservative approach to kidney surgery, in which doctors sought to remove the stone while sparing the kidney. In the late 1880s, surgeons began experimenting with the partial nephrectomy, removing just part of the affected kidney.

Nephrolithotomy gained popularity in the late 1800s, but technology at the time still lagged behind. For one thing, doctors could never tell for sure whether the patient was suffering from a stone or some other urinary tract disorder because they had no means with which to visualize the urinary system. Doctors relied on manual palpation or a needle probe of the region, both of which were unreliable. In many cases, patients underwent operations only to have the surgeon find nothing wrong.

But there were also substantial gains made in the field of surgery during the nineteenth century. With the introduction of anesthesia, patients no longer had to suffer a painful ordeal during surgery. The invention of antiseptic prevented deadly infections, and the suture controlled bleeding during surgery. Endoscopes allowed doctors for the first time to see inside a patient's urinary tract to diagnose the problem. But probably the most significant diagnostic tool to date was radiography. Discovered in the late 1800s, x-rays allowed doctors to visualize the entire urinary tract (see Chapter 4). At first, the technique was primarily used to diagnose stones, but it eventually took on a wide range of applications. In 1929, the introduction of the **intravenous pyelogram (IVP)**, an x-ray using a special dye that clearly reveals the urinary tract, revolutionized the diagnosis and treatment of urinary stones.

The next big innovation in stone surgery was the art of **lithotripsy**, or breaking up stones for easier removal. Although the technique did not catch on until the 1800s, the idea of lithotripsy was first described by Celsus (c. 25 BCE–50 CE) using a scoop, chisel, and hammer. The Arab physician Rhazes detailed a stone-crushing technique in the tenth volume of his work, *Continens*, in which a pincer was used to break away pieces of the stone until it was small enough for removal. In the 1800s, surgeons began experimenting with several instruments that could be threaded up the patient's urethra and into the bladder or kidneys to crush the stone. Fournier de Lempdes (1783–1848) devised a cage-like instrument, which he called the litholepte. As the doctor rotated the instrument, five metal blades opened and closed and a file inside wore away at the stone.

But one problem still remained—how to remove the stone. In 1818, French surgeon Jean Civiale (1792–1867) designed a three-bladed instrument called a trilabe. When the trilabe was inserted into the patient's bladder, the blades opened out and closed around the stone. A burr inside the blades rotated to crush the stone and catch the fragments for removal. The trilabe was never used on a patient, but Civiale's next invention, the lithontripteur, was successfully used on some 1,600 patients. The instrument grasped, crushed, and removed the stone with a two-bladed forceps. Leroy d'Etoilles introduced a similar instrument in 1822, called the lithoprione. It, too, caught the stone between metal blades and crushed it with a rotating burr.

The first to introduce the concept of modern lithotripsy was Baron L. S. Heurteloup (1793–1864). His percussion lithotrite first caught the stone between two grooved blades, then a vise held the stone in place as an external hammer drove the blade into it. He called his technique "lithotripsy."

Patients welcomed the new technique as a less painful and less invasive alternative to surgery, and it slowly gained acceptance in the late 1800s. Over the next several years, several modifications were made to the techniques and equipment involved in the procedure. Physicians attempted several different methods of lithotripsy. The one used by Civiale crushed the stones over a series of procedures, allowing the fragments to pass between sessions. Other doctors performed the procedure only once, but repeatedly struck the stone to break it up completely. Up to this point, French surgeons were the most prolific practitioners of lithotripsy, but an American was to make one of the most significant improvements to the technique. Boston urologist Henry J. Bigelow (1818–1890) modified Civiale's lithotrite to crush larger stones, wash out the particles, and catch them via a rubber bulb and glass trap. He called his procedure litholapaxy, or "Bigelow's Operation."

Stone surgery had come a long way by the late 1800s, but it had even further to travel, for doctors were still performing the procedure blind, unable to see the stone they were attempting to crush. The invention of Nitze's cystoscope (discussed later in this chapter) in 1877 allowed urinary surgeons for the first time to visually follow the course of their work. In 1908, the "father of American urology," Hugh Young (1897–1941), introduced the first lithotrite with an optic system.

In the early 1950s, researchers began experimenting with new ways to fragment and remove stones, using ultrasonic energy rather than blades and hammers. In this technique, high-frequency energy was transmitted through a metal probe to crush the stone. In the 1960s, lasers were used for the first time to destroy stones. Since the mid-1980s, extracorporeal shock wave lithotripsy (ESWL) (see Chapter 4) has become the preferred method for removing stones. ESWL crushes stones with a powerful electric current, which is applied to the outside of the body over the location of the stone.

KIDNEY REMOVAL AND TRANSPLANTS

Kidney surgery dates back to the time of Hippocrates, even though the "father of medicine" disdained the practice. He permitted surgery only in cases where the kidney had become infected and an abcess had to be drained. Kidney transplantation or removal would be impossible for nearly 2,000 more years, because doctors feared that the procedure would most likely result in uncontrollable bleeding and infection, not to mention unbearable pain to an unanesthetized patient.

The era of modern kidney surgery did not truly begin until the end of the 1800s, with the introduction of anesthesia, although several attempts were made to raise interest in the procedure prior to that. In the 1600s, Stephen Blankaart (1650–1702) suggested that kidney removal could help patients suffering from kidney stones. But other surgeons of the day scoffed at the idea that humans could tolerate such a procedure.

The first nephrectomy was performed, albeit accidentally, by Erastus Bradley Wolcott (1804–1880), Wisconsin surgeon general during the Civil War. As Wolcott operated to remove what he thought was a cystic tumor of the liver, he mistakenly removed the patient's kidney, which was attached to the 2.5-pound tumor. The patient died within fifteen days from "exhaustion caused by infection." Less than a decade later, German surgeon Gustav Simon (1824–1876) (see the section on lithotomy earlier in this chapter) performed the first removal of a healthy kidney. The operation lasted for forty minutes, during which time the female patient was sedated with chloroform. The woman suffered a severe infection that took several months to heal, but she lived to prove that a patient could survive with just one kidney.

Nephrectomy didn't catch on immediately, although several other surgeons attempted the operation, including Simon. But despite his initial success, the overall survival rate remained low—about 20 percent—and doctors seriously questioned whether nephrectomy had any real merit. But with the advent of antiseptic to prevent infection, and improved methods of diagnosis aided by Nitze's cystoscope and Roentgen's x-ray, the operation began to gain acceptance in the late 1800s.

Next, surgeons wanted to learn not only how to remove a diseased kidney but also how to replace it with a healthy organ. But the problem they had to overcome seemed almost insurmountable: rejection. The human body, when presented with foreign tissue, launches an all-out assault that results in the rejection of that tissue. The first surgeons to attempt kidney transplants found this out very quickly. Actually, initial attempts at kidney transplantation were not attempted on humans, but on animals. In 1902, Austrian surgeon Emerich Ullman (1861–1937) experimented with kidney transplantation on dogs. In every case, the kidneys were rejected within a few days. That same year he attempted, unsuccessfully, to transplant a pig's kidney into a woman with end-stage renal disease. In 1906, French surgeon Mathieu Jaboulay (1860–1913) was also unsuccessful when he attempted to graft a pig's kidney to the blood vessels in the arm of a patient with chronic kidney failure.

In the early 1900s, Jaboulay's student, Alexis Carrel (1873–1944), pioneered many of the transplantation techniques still in use today. He realized that cadavers were viable sources of organs, and he introduced cooling as a means to preserve organs for transplantation. And, when his attempts at autografting dog kidneys failed, he suggested that the "principle of immunity" might be to blame.

The first human-to-human transplant was performed in 1933 by a Ukranian surgeon named Yu Yu Voronoy. The kidney was harvested from a cadaver—a patient who had died from a head injury several hours earlier—and transplanted into a patient suffering from mercury poisoning. Unfortunately, the donor and recipient's blood types were incompatible, and the patient died just days later. Voronoy subsequently performed five more transplants, all without success. It was not until the early 1940s that doctors began to understand the "principle of immunity" Carrel had proposed.

During World War II, a British immunologist named Peter Medawar (1915–1987) and his colleague Thomas Gibson were experimenting with skin grafts on severely burned pilots, when they noticed that the first set of grafts survived for several days longer than the second set, although both had been transplanted from the same donor. The researchers surmised that the patient developed an immune response to the foreign tissue, which led to the rejection of the skin. This discovery would pave the way for the first successful organ transplant.

In the 1940s and 1950s, doctors attempted several transplants using cadaver tissue. All achieved varying degrees of success, but ultimately all of the organs were rejected. To avoid the problem of rejection (because doctors did not know in those days how to stifle the immune response), doctors attempted transplants between identical twins, who shared the same genes. In 1954, Joseph Murray (1919–) performed the first successful kidney transplant on a 23-year-old identical twin who was dying from kidney failure. The patient lived for twenty years. By 1958, Murray and his colleagues had reportedly performed eight kidney transplants among sets of identical twins.

But Murray's success, although inspiring, did not overcome the primary hurdle in transplant technology—how to suppress the immune response so the patient would not reject the foreign tissue. In the late 1950s, Boston physicians Robert Schwartz and William Dameshek discovered that immunosuppressive drugs could be used for this purpose. At first, immunosuppression was achieved by whole-body irradiation. Radiation did prevent rejection, but it often killed the patient it was attempting to save. Subsequently, doctors tried immunosuppressive drugs such as prednisone, 6-mercaptopurine, actinomycin C, and imuran, but the rate of failure after kidney transplantation remained high. In the early 1960s, Joseph Murray and his colleagues successfully used the drug azathioprine to prevent rejection in their kidney transplant patients. A combination therapy of azathioprine and steroids soon became a staple of antirejection therapy.

In 1960, Peter Medawar made the next great stride in transplantation technology by introducing the idea of tissue typing. By comparing genetic markers, called human leukocyte antigens (HLA), doctors would be able to determine whether the donor and recipient were a likely match before attempting surgery (see Chapter 5).

But by far the most dramatic revolution in transplantation surgery came with the introduction of the immunosuppressive drug cyclosporine in 1972. The drug was more effective than its predecessors in preventing rejection, without harmful side effects. When the U.S. Food and Drug Administration (FDA) approved cyclosporine for general use in 1983, transplant survival rates began to skyrocket. During the 1980s, doctors furthered their understanding of the rejection process and learned how to more accurately match organ donors and recipients. Several new immunosuppressant drugs were added to the antirejection arsenal in the 1990s, including mycophenolate mofetil (CellCept) and a new formulation of cyclosporine called Neorol.

By the turn of the twenty-first century, advances in transplantation technology had led to a dramatic rise in patient survival rates. In the 1970s, a patient who received a cadaver kidney had only a 50 percent chance of surviving one year. In 2002, that same patient had more than an 80 percent chance of surviving five years. But even as survival rates improved, doctors were faced with another challenge: how to meet the growing demand for organs. As of 2002, there were more than 50,000 people on the United Network for Organ Sharing waiting list for a kidney, and that number was on the rise. Organ shortages forced scientists to turn their attention to other potential organ donors—including baboons, genetically altered pigs, and other animals. While **xenotransplantation**, as it is called, has the potential to eventually solve the organ crisis, it is still years away from practicality, because doctors have yet to figure out how to overcome the problem of animal-to-human tissue rejection (see Chapter 5).

DIALYSIS

Kidney transplant surgery didn't become truly successful until the latter part of the twentieth century, but prior to that time, doctors still had to grapple with the problem of sustaining patients whose kidneys had failed. The challenge was to remove the wastes the kidneys would normally filter out of the body. In the late 1800s, the idea first emerged to design a machine to cleanse the blood. This was no easy task, considering the sophistication of kidney architecture and the complicated nature of its filtration system.

In the mid-1800s, Scottish chemist Thomas Graham (1805–1869) was experimenting with molecular separation, using vegetable parchment coated with albumin as a selectively permeable membrane (a membrane that only allows certain types of molecules to pass through). With this method, he was able to filter urea out of urine. He called his technique **dialysis**.

John Jacob Abel (1857–1938) constructed the first filtering membrane in 1898. In 1913, he and his colleagues at Johns Hopkins University constructed an apparatus for removing toxins from the blood, which is considered the forerunner of the artificial kidney. Their technique, which they called "vividiffusion," circulated blood through a series of tubes made of celloidin. To prevent the blood from clotting as it was filtered, they used hirudin, a substance produced by leeches, as an anticoagulant. Once filtered, the blood was returned to the animals' vein. The inventors wrote, "This apparatus might be applied to human beings suffering from certain toxic states, especially if due to kidney damage, in the hope of tiding a patient over a dangerous chemical emergency." But the device was never used on humans.

The first human dialysis was performed in 1924 by German doctor George Haas (1886–1971) on a patient with acute kidney failure. The patient, who was suffering from terminal uremia, was treated for 15 minutes without complication. But it wasn't until World War II that the first artificial kidney was created. Willem Johan Kolff (1912–), a Dutch-born American physician, constructed the first rotating drum artificial kidney. With his device, the patient's blood entered cellulose tubing that was wound around a rotating drum inside a tank filled with dialysate solution. As the drum was hand cranked, the patient's blood was propelled through the tubing and filtered. Finally, the blood was pumped back into the patient's vein. Kolff used his artificial kidney on a patient for the first time in 1943. In that same decade, the anticoagulant heparin was introduced, replacing the unsafe hirudin.

After World War II, Kolff moved to the United States, bringing several of his machines with him. Over the following decade, he developed the next generation of dialysis machines called "twin coil" **dialyzers**. These machines, which used disposable tubing for filtration, were instrumental in saving soldiers who suffered kidney injuries during the Korean War.

During the 1960s, dialysis underwent a number of improvements. Scientists developed better access routes into the body, called fistulas and **catheters**. A fistula is an entryway through which the dialysis machine can be connected to the patient's body. A catheter is inserted into a vein in the patient's neck, chest, or leg to provide quick access for dialysis. In the following decade, a new law required Medicare to pay for 80 percent of dialysis treatment costs, paving the way for new dialysis medical centers and more widespread patient use. In the 1980s and 1990s, the development of smaller and more efficient dialysis machines, as well as improved monitoring and safety devices, allowed patients on dialysis to travel and enjoy normal lives.

In a single century, the artificial kidney progressed from a crude membrane to a highly sophisticated filter. But as far as the dialysis machine has come, its days may be numbered. Scientists have now created a bio-artificial kidney, lined with living kidney cells that reabsorb vital electrolytes, water, and glucose normally removed from the blood during filtration. Although the invention is still in the early stages of development, researchers hope to one day be able to implant the device and have it take over all the functions of a normal kidney. When that day arrives, the dialysis machine may become obsolete.

PROSTATE TREATMENT AND SURGERY

The earliest physicians understood little about the mysterious prostate gland, why it swelled, and how it inhibited the flow of urine. But, then, they had little need to understand the prostate. Most men did not live into their 50s or 60s, when prostate ailments are most likely to develop. Furthermore, thousands of years ago doctors gleaned much of their anatomical understanding from animal dissection, which provided a poor reference for the human prostate.

The first description of the prostate is found in the writings of the great anatomist, Herophilus of Chalcedon. Rufus of Ephesus (98–117) also described the gland, which he called the *parastatus glandulus*, meaning "standing before." He is credited with giving the prostate its name. The French anatomist Jean Riolan (1580–1657) was the first to surmise that the prostate might play some role in obstruction when he noted that "the neck of the bladder may be obstructed by a tumour of the prostate gland." In fact, most doctors at the time thought that prostatic enlargement was caused by a cancerous tumor. Other explanations ranged from urinary stones to congested blood flow and venereal disease. Even prolonged sitting and excessive horseback riding were implicated. But the first real understanding of what we now know as benign prostatic hyperplasia (BPH; see Chapter 7) was set forth by Giovanni Morgagni (1682–1771), an Italian professor of anatomy. Morgagni observed that prostate enlargement inhibited the flow of urine in his elderly male patients. He referred to the swollen middle lobe of the prostate as the "caruncle," and for many years thereafter, the prostate was referred to as "Morgagni's caruncle."

Although the prostate's role in urinary obstruction was not fully understood until the eighteenth century and wasn't effectively treated until the twentieth century, early doctors were able to relieve blockages with a number of surgical and nonsurgical techniques. They used hollow rods dipped in caustic solutions to burn away the blockage, or forced a sharp-pointed instrument up through the center of the prostate to bore a hole to the bladder. After the instrument was removed, the patient could urinate through the new route. The only trouble was that the patient was often unable to urinate with any degree of control.

In 1786, a Scottish anatomist and surgeon named John Hunter (1728–1793) discovered that **orchiectomy**, or castration (removal of the testicles), halted prostate growth in young male animals. Castration was attempted many times in the 1800s, but the high mortality rate eventually forced doctors to abandon the technique. During the same century, doctors tested a variety of substances to shrink prostatic obstruction. They injected iodine, silver nitrate, and even cocaine into the gland. But severe infection usually resulted. Drugs were also given orally to shrink the prostate and relieve symptoms. These included hemlock, mercury, hydrochlorate of ammonia, iodine, and bromine.

None of these early techniques achieved much success in reducing prostate size and eliminating the urinary blockage. Before the advent of prostate surgery, doctors could do little to relieve prostate symptoms, and most patients were forced to endure a lifetime of daily catheterization. It wasn't until the mid-1800s that doctors first attempted direct surgery to relieve a prostatic blockage. The first of such surgeries was probably unintentional: the surgeon cutting into the prostate in a mistaken attempt to reach a stone. One such unknowing attempt was made by surgeon Joseph Covillard, who in 1639 removed a wedge of prostate tissue via a perineal approach (between the scrotum and rectum). Most of the early prostate surgeries were in fact attempted through the perineal approach, which was well known at the time from stone surgery.

In 1886, W. T. Belfield, a pioneer in the field of prostate surgery, performed the first deliberate operation to remove obstructing tissue. The following year, Arthur Ferguson McGill, a British surgeon, performed a partial **prostatectomy** (partial removal of the prostate) through the suprapubic approach (an incision made above the pubic bone) to remove what he called a "tumor at the base of the bladder." When he realized that the offending tissue was actually from an enlarged prostate, he made the decision to always operate on his patients when the prostate blocked the bladder entrance. McGill reported that he had operated on three BPH patients in 1887, using scissors and forceps to remove the obstructing tissue. All three patients recovered and were subsequently able to urinate on their own.

By the end of the nineteenth century, surgery to remove enlarged prostate tissue had become more routine. In 1890, Dr. Belfield reviewed all previously published cases of prostate surgery, and found that most patients—about 68 percent—were able to urinate on their own following surgery. But because the risk of infection was still very real, most doctors reserved the procedure for their sickest patients. When they did perform the surgery, surgeons opted to remove only part of the prostate (partial prostatectomy) to preserve continence and reduce the risk of complications.

In 1900, an Irish surgeon named Peter Freyer (1852–1921) performed the first total suprapubic prostatectomy at St. Peter's Hospital in England. Seven years after the operation, the patient was in perfect health and able to control urination. Throughout his career, Freyer performed more than 1,600 of these surgeries with a mortality rate just over 5 percent. His work became so well known in Europe that the technique he perfected became known as "Freyer's operation." In 1902, Hugh Young revived the perineal approach, and it remains the technique of choice among surgeons today.

The early 1900s saw the introduction of electrotherapies to burn away obstructing prostate tissue. Using a high-frequency current applied through a cystoscope, doctors could remove tissue with less blood loss and shorter recovery time than open prostate surgery. In 1926, Maximilian Stern of New York invented a movable tungsten loop that could cut away cylinder-shaped pieces of prostate tissue. His device was the predecessor of today's **transurethral resection of the prostate** (**TURP**; see Chapter 7) procedure. South Carolina physician Theodore M. Davis improved upon Stern's invention in 1931, perfecting the electric current supply and widening the loop to remove larger pieces of tissue, thereby preventing uncontrolled bleeding. In that same year, Joseph McCarthy and his colleague developed the Stern-McCarthy resectoscope. The instrument was outfitted with a new lens system that vastly widened the surgeon's field of vision, as well as a wider loop and sharper cutting instrument that allowed doctors to precisely determine how much tissue to remove and to better control bleeding. In the 1950s, fiber optics made the flexible endoscope possible, allowing doctors to get a better view while they operated.

Over the next several decades, doctors added medications to their arsenal of prostate reduction weapons. **Alpha adrenergic blockers** and **5 alpha-reductase inhibitors** were—and still are—used to shrink the prostate and release the urinary obstruction without the need for surgery (see Chapter 7). New surgical techniques that use microwave, radio-frequency, or laser energy to destroy prostate tissue have reduced recovery times without compromising treatment quality. But despite these advances, a prevention or cure for BPH will remain elusive until the medical community can determine what causes the mysterious affliction.

CANCERS OF THE URINARY SYSTEM

Although it was not until the twentieth century that doctors understood how or why cancerous tumors formed, many physicians were able to identify abnormal growths while performing kidney and bladder operations. Hippocrates exhibited a relatively astute understanding of the malignant nature of these growths when he said, "Blood or pus in the urine indicated ulceration either of the kidney or the bladder." In 1673, a surgeon named Segerus described the enlargement in a woman's right kidney as a "cancer." French surgeon François Chopart (1743–1795) in 1791 distinguished between tumors in the bladder and bladder neck. He said urinary bleeding might be due to tumors, and he understood that benign tumors could turn cancerous.

The first kidney tumor was described in 1810. Prostate tumors were a bit more difficult to identify, as little was known about the gland. In 1832, English surgeon Sir Benjamin Brodie (1783–1862) distinguished prostate cancer from benign prostate growth when he referred to the patient's prostate as "not much enlarged but of a stony hardness."

Over the years, surgeons tried a number of different methods to remove tumors from the bladder, the kidneys, and later the prostate. They tore, crushed, scraped, and tied off these lesions, often with deadly results. Some lithotomists, including Pierre Franco, attempted to remove tumors while they were operating to extract stones. During the 1800s, doctors began to use instruments designed to grasp stones—such as the trilabe (Jean Civiale's three-bladed instrument) and forceps—to remove tumors.

Cancer of the Kidney

The first nephrectomy to remove a kidney tumor was performed by Erastus Wolcott in 1861. The tumor, which is said to have weighed over two pounds, was believed at the time of surgery to be a liver cyst, but was later discovered to be a kidney cancer. In 1887, Vincenz von Czerny (1842–1916) performed the first partial nephrectomy, removing the part of a patient's kidney that housed the cancerous tumor. For most of the twentieth century, doctors opted to remove the entire kidney (radical nephrectomy) whenever cancer was present. But by the end of the century, they began to opt for partial nephrectomy in an effort to spare kidney function.

The introduction of intravenous pyelography (IVP) in 1929 revolutionized the diagnosis of both kidney and bladder cancers, and it remains the standard today. By injecting an iodine-containing contrast dye into a patient's blood vessel, doctors can now view an image of the entire urinary tract on x-ray film and pinpoint the location of tumors and other abnormal masses.

Cancer of the Bladder

At the end of the nineteenth century, surgeons for the first time attempted to remove cancerous bladder tumors. The German surgeon Theodor Billroth (1829–1894) introduced the modern era of bladder surgery in 1874, when he documented the removal of a large bladder tumor through a suprapubic (making an incision above the pubic bone) approach. Other surgeons quickly began following his lead. In 1884, Eduard Sonnenberg reported performing the first partial **cystectomy** (bladder removal). Three years later, Bernard Bardenheuer of Germany performed the first total cystectomy for bladder cancer.

The problem with removing the entire bladder, however, was urinary diversion: how to reroute and store urine without aid of the bladder. Bardenheuer attempted to implant a patient's ureters in their bowel, a procedure he had tried successfully on animals. But he mistakenly left the ureters in the pelvis, and the patient died two weeks later. Other doctors tried a number of different methods, including suturing the ureters to the urethra—but most were unsuccessful. It wasn't until the latter part of the twentieth century that doctors discovered how to create an artificial urinary diversion (called an **ileal conduit** or loop) using a piece of the patient's intestine, and figured out how to remove the urine from the patient's body by way of a **stoma** bag or catheter (see Chapter 6). A newer technique uses part of the intestine to create a new bladder, which is then attached to the patient's urethra.

In the early 1900s, electrotherapies were introduced to burn away, rather than cut away, cancerous bladder tissue. In 1910, New York urologist Edwin Beer burned off two bladder tumors with a high-frequency current applied through a cystoscope. The procedure produced less bleeding than traditional surgery and eliminated the need for hospitalization. By 1914, the procedure had gained popularity, and it remains in use today (see Chapter 6).

Around the same time, doctors were also investigating whether introducing chemicals into the bladder could destroy cancerous tumors. Doctors tried everything from boiling water to silver nitrate, but none worked. In the early 1900s, they turned their attention to radioactive compounds. The first attempts at instilling radium into the bladder cavity achieved little success. But by the end of the twentieth century, doctors had mastered radiation therapy and had at their disposal a wide variety of techniques, from chemotherapy to radiotherapy. They are even beginning to learn how to teach the patient's own defense system to kill cancer cells (an approach called **immunotherapy**).

One of the greatest advancements to the treatment of bladder cancer was the introduction of the cystoscope, a long thin scope that has enabled doctors to examine tumors and remove a slice of tissue for biopsy before performing open surgery. The cystoscope revolutionized the diagnosis of bladder cancers because it allowed doctors to learn how far and how fast the cancer was progressing. The first to observe and remove a bladder tumor through a cystoscope was Josef Grünfeld in 1885. By 1905, he had reported using the cystoscope in more than 270 successful operations. The inventor of the electrically lit cystoscope, German urologist Max Nitze (1848–1906), published a paper on the cystoscopic diagnosis of bladder tumors in 1889, and the technique gradually gained popularity. Nitze used his cystoscope to cut off bladder tumors and burn them at their base. He was also the first to differentiate between benign and malignant bladder tumors, overturning the previously held belief that all tumors were cancerous. Doctors still use **cystoscopy** to view the inside of the patient's bladder and remove cells for biopsy. That tissue is then examined under a microscope by a pathologist, who can determine whether it is indeed cancer and, if so, how far it has spread. In the 1940s, doctors added **cytology**—examining the patient's urine under a microscope for cancerous cells—as yet another diagnostic tool. Together, IVP, cystoscopy, and cytology provide modern doctors with powerful tools for grading and staging cancer progression.

Cancer of the Prostate

Surgeons struggled with the diagnosis and treatment of prostate tumors throughout the late 1800s and early 1900s. In 1867, Theodor Billroth performed the first partial perineal prostatectomy (entering through the area that lies between the scrotum and anus in men) to remove a tumor the size of a duck's egg from a 30-year-old patient. A year later, the cancer returned and the patient died. Perineal prostate operations attempted in the late 1800s were rarely successful. In 1883, the first total perineal prostatectomy for cancer resulted in the patient's death two weeks later. But the method was not abandoned. Around the turn of the century, it was adopted by Hugh Young. Young had initially performed prostatectomies through the suprapubic approach because of its speed and ease. But, encouraged by his success in using the perineal approach on BPH patients, he soon switched to this method for his cancer patients. He made several improvements to the procedure and invented an instrument that expedited prostate removal, called the "Young punch." Thanks to Young's work, the perineal route became the preferred access for prostate removal surgery.

The technique was effective, but it wasn't without side effects. When doctors removed the prostate, they also removed or destroyed the nerves responsible for sexual function, thus rendering the patient impotent. In the early 1980s, Patrick C. Walsh designed a new approach to radical prostatectomy, which removed the cancerous prostate while sparing the nearby nerves to preserve urinary and sexual function. The operation, called nerve-sparing radical retropubic prostatectomy, remains the standard in prostate cancer surgery. Modern surgeons also have at their disposal laparoscopic prostatectomy, in which tiny instruments are inserted through the patient's abdomen to remove the cancerous tissue. This technique also preserves sexual function, with less blood loss and shorter recovery time than open prostate surgery.

While advances were being made in the field of prostate surgery, researchers were looking into nonsurgical treatment options that could replace the knife and operating room. In the 1940s, Charles Huggins (1901–1997), a Canadian urologist, proved that prostate cancer could not grow on its own but was dependent on male hormones for survival. He determined that by blocking male hormones with the female hormone estrogen, or by removing the testicles, he could halt the progression of prostate cancer. His discovery led to the development of several hormone-based therapies, which are still used to treat prostate cancer patients.

URINARY DEVICES

To diagnose urinary stones, blockages, and disease, doctors at first were forced to rely on external clues. They would examine the urine for sediment or color changes, and palpate the body where the kidneys and bladder lay to feel for hard masses. In terms of treatment, doctors could do virtually nothing without making an incision, all the while hoping that their hunch was correct and that their patient would not die from blood loss or infection. But thanks to new tools—the cystoscope, x-ray, and catheter—developed during the late nineteenth and early twentieth centuries, doctors were able for the first time to noninvasively diagnose and treat urinary disease.

The cystoscope opened up and illuminated the interior of the bladder to help doctors stage and grade cancerous tumors, remove tissue for biopsy, and identify stones. The x-ray revealed the complicated inner workings of the urinary system and displayed abnormalities for easier diagnosis. And the catheter reopened the flow of urine to save patients who would have otherwise died from a blockage.

Endoscopy: The Birth of the Cystoscope

From the earliest days of medicine, doctors were somewhat capable of identifying and treating diseases of the urinary system. They were able to point out changes in the urine, as well as obvious swelling and other hard masses in the kidneys and bladder. However, they were helpless to render an accurate diagnosis and fully treat the patient without being able to see—and carry their instruments—into the dark cavities of the body. Then came the birth of endoscopy. Endoscopy comes from the Greek words *evdov*, meaning within, and *okonelv*, to examine. Hippocrates made the earliest mention of endoscopy when he described a speculum for rectal examination. Roman physicians were believed to have used simple specula to examine the urinary organs.

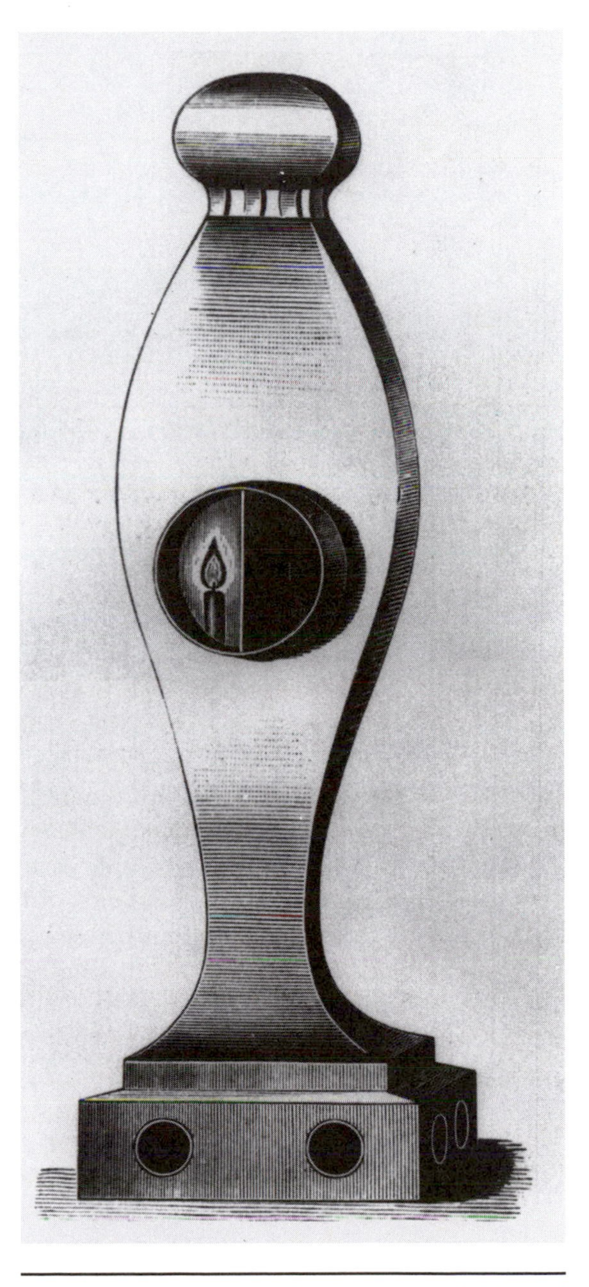

Bozzini's early cystoscope. Wood engraving. © National Library of Medicine.

But modern endoscopy did not make its debut until the nineteenth century, when Philip Bozzini (1773–1809) realized the necessity of introducing a light source to illuminate the interior of the body. Bozzini's first endoscope (see illustration) was a crude device consisting of a tin tube encased in cardboard and leather, to which was attached a lit candle on a stand. A reflector directed the candlelight down the funnel to illuminate the organs. Bozzini called his device the lichtleiter. The trouble was that the single candle barely shed enough light for a doctor to work and the heat from its wick often burned the patient.

In 1826, Pierre Ségalas (1792–1875) released a modified version of Bozzini's lichtleiter, which he called the urethrocystic speculum. The device was made of two silver tubes, two metal mirrors, and two lit candles. One of the tubes was darkened to shade the doctor's eye from outside light. The mirrors projected the candlelight along the tube to illuminate the area while the doctor worked. Ségalas used his speculum primarily to examine the urethra for

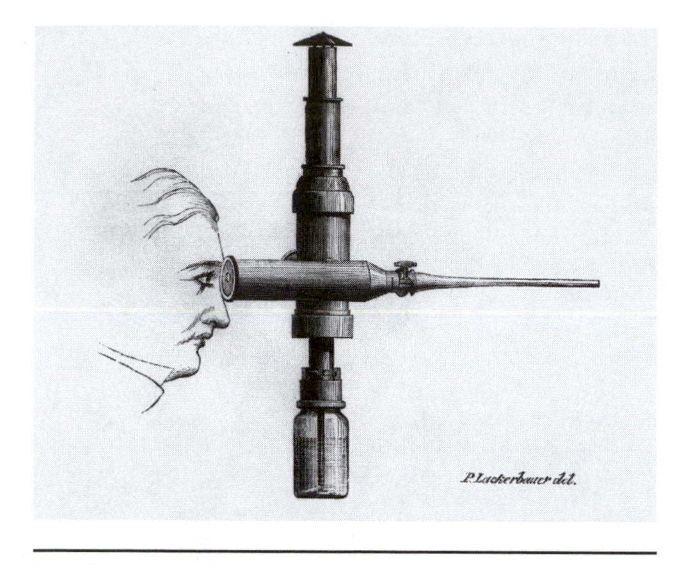

Jean Desormeaux's cystoscope, 1853. Wood engraving. © National Library of Medicine.

stones and other lesions. Although the device shed slightly more light than Bozzini's version, it still had not overcome the problem of placing a hot candle so close to a patient's sensitive skin.

The lichtleiter was improved yet again by the "father of endoscopy," Antoine Jean Desormeaux (1815–1881) (see illustration). In 1853, the surgeon introduced his own version of the device to the French Academy of Medicine. Like its predecessors, the instrument consisted of a long tube with lenses and a mirror. The light in this case was provided by a wick burning a mixture of alcohol and turpentine. The flame was brighter and more contained than in previous endoscopes, but could still burn the patient. The lenses focused light on the mirror, which reflected that light down the catheter. Desormeaux's endoscope was used by urologists in Europe and America.

The field of endoscopy took a great step forward in 1877, when German urologist Max Nitze (1848–1906) produced the first electrically lit cystoscope. Nitze realized that he needed to bring light into the urinary tract rather than try to illuminate it from the outside. As he said, "In order to light up a room one must carry a lamp into it." After several failed experiments, he finally hit upon a light source that worked—a heated platinum wire loop fixed at the end of the cystoscope. Nitze's lens system also improved upon its predecessors by widening the physician's field of vision. But the system was still bulky and the wire continually threatened to either burn out or burn the patient. In 1879, Nitze collaborated with an instrument maker named Josef Leiter in Vienna to develop the Nitze-Leiter cystoscope. The wire loop was still the source of light, but it was placed at the tip for better illumination and water-cooled to protect the patient from burns.

Then came an innovation that forever changed the course of endoscopy: the invention of Thomas Edison's electric light bulb in 1879. No longer would the cystoscope need to rely on a hot and bulky light source. In the late 1880s, Nitze's original accomplice, Leiter, fitted the cystoscope with Edison's electric light (see illustration). By the early twentieth century, new cystoscopes were outfitted with a small but brighter "low-amp" light, improved lens systems that widened the field of vision, and a telescope that allowed doctors to view the urinary organs right side up.

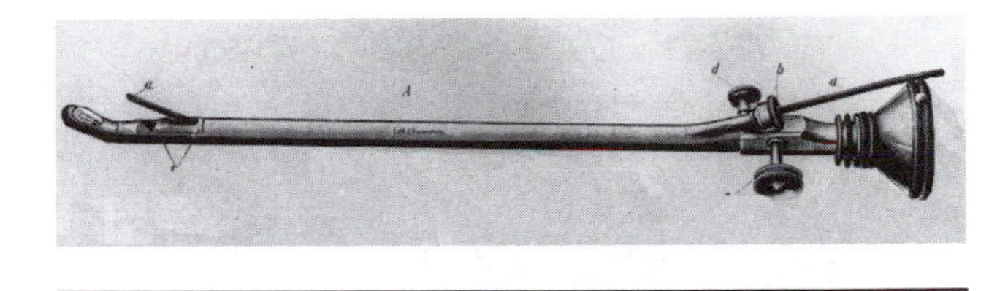

Nitze's modified cystoscope with Edison's electric light bulb. © National Library of Medicine.

Scientists at the turn of the century were also concentrating on developing a better operating cystoscope, equipped with forceps and other surgical tools. In 1906, Bransford Lewis debuted his Universal Scope, the precursor to the modern cystoscope. The device had windows on both convex and concave surfaces, allowing for illumination and observation on both sides. It was also equipped with forceps, scissors, and dilators for operating on the patient. By the end of the century, the cystoscope had become thin enough to easily traverse the urinary system. A tiny camera was added to give surgeons a multiangled view of the bladder. And tiny instruments inserted through the cystoscope allowed doctors to easily remove tissue for biopsy.

The X-ray

In 1895, a German engineer named Wilhelm Conrad Roentgen (1845–1923) was experimenting with cathode rays when he discovered a new type of ray—the x-ray (see photo). When objects—or people—were placed in the path of this ray, they became transparent when recorded on a photographic plate. Physicians embraced Roentgen's discovery, adopting the x-ray as a diagnostic tool. In 1896, Scottish physician John MacIntyre (1857–1928) was the first to find a kidney stone with an x-ray. The exposure took a full twelve minutes, but surgery confirmed that the x-ray had been correct.

But although the x-ray worked well for viewing the skeletal structure, and it could reveal some large stones and other abnormalities of the urinary tract, it was not able to provide a clear enough image of the organs to be of real use to urologists. A greater contrast would be needed to visualize the kidneys, bladder, and ureters on film. In 1906, Fritz Voelcker (1872–1955) and Alexander von Lichtenberg (1880–1949) of Germany unintentionally discovered a better method for viewing the urinary tract, called pyelography. The two instilled a silver solution called collargol into the bladder and found that it outlined part of the urinary system. Several years down the road, researchers began experimenting with a number of opaque solutions, including more concentrated silver solutions, which were often toxic to patients. The challenge was to create a nontoxic solution that was radiopaque (absorbs radiation) enough to be picked up by an x-ray machine. In 1923, doctors at the Mayo Clinic in Rochester, Minnesota, tried oral and intra-

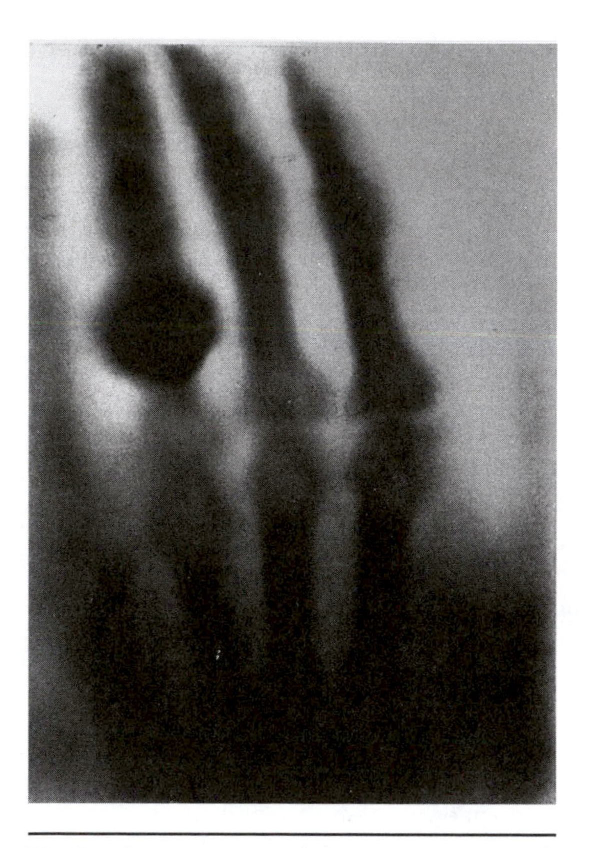

The first Roentgen x-ray, of Mrs. Roentgen's hand. © National Library of Medicine.

venous injections of sodium iodide. The solution revealed a hazy outline of the urinary system.

In 1929, a New York doctor named Moses Swick began investigating with an organic iodide compound called uroselectan. The compound had been used in Germany for treating animal infections, but Swick was aware of its radiopaque properties. He injected the material into a patient's vein, taking x-rays as the material moved through the urinary tract. In large concentrations, uroselectan caused headaches, nausea, and double vision, but a nontoxic version of the drug worked well with fewer side effects. Swick's discovery led to the modern version of intravenous pyelography.

Over the next few decades, researchers devised faster, more accurate noninvasive diagnostic tools, like **ultrasound**, **computed tomography (CT)**, and **magnetic resonance imaging (MRI)**. In the 1950s, doctors adopted ultrasound technology—the same high-frequency sound waves used to detect submarines—to reveal the interior of the abdomen and urinary tract. In 1972, CT scanning allowed for higher-dimensional imaging by creating a cross-sectional view of urinary organs. And a decade later, the MRI further enhanced imaging techniques by giving doctors the ability to distinguish normal from diseased tissue.

The Catheter

Urinary retention from stones or prostate enlargement undoubtedly plagued the ancient Egyptians, Greeks, and Romans, just as it does modern humans. But without the benefit of modern hospitals and surgical techniques, doctors in those eras were forced to treat the condition with whatever rudimentary tools were at their disposal. Ancient texts describe the first catheters as made of reeds, straws, and palm leaves. But these delicate instruments were likely to break inside a patient, and doctors began searching for more stable and reliable tools.

The Greeks and Romans fashioned their catheters out of copper, bronze, or tin, with a single opening at the end. At the beginning of the Christian era, Aulus Cornelius Celsus described the technique by which the Romans used their catheters:

> The patient is to be placed on his back, as described for operation at the anus, upon a stool or couch. The physician standing at his right should, in the male subjects, lay hold of the penis with his left hand, while with his right he passes the catheter into the urethra; when he has reached the neck of the bladder, he is to give the instrument, together with the penis, a slight inclination downwards, and to push it on into the bladder itself, and to withdraw it after the urine has been evacuated.

The trouble with a metal catheter was its rigidity. Catheters made of bronze or copper were difficult for the physician to manipulate and painful for the patient to tolerate. Arabic doctors were believed to have been the first to address the problem by devising a flexible catheter made of animal skins. The catheter was treated with a salve of white lead and ox blood and lubricated with soft cheese. The base of the catheter was made of silver, chosen for its malleability and supposed antiseptic properties. Doctors continued to use silver catheters, with few modifications, for the next several centuries.

In the 1700s, silver and woven silk catheters started to come into use. The catheters were first soaked in linseed oil and then dried. Silk was soft and malleable, but it broke easily and pieces often became lost in the patient's body. In 1839, a bankrupt hardware merchant from Philadelphia named Charles Goodyear unintentionally revolutionized catheter design when he discovered the vulcanization process that created rubber. The new rubber catheters were soft, durable, and flexible.

In 1930, F.E.B. Foley, a doctor in St. Paul, Minnesota, made the next great innovation when he developed a catheter made out of soft latex rubber. The catheter was attached to an inflating tube and balloon that held the device in place while it continuously drained urine from the patient's body. Other doctors came up with similar designs, but Foley's device was picked up by the Anode Company, and his name has been attached to it ever since.

4

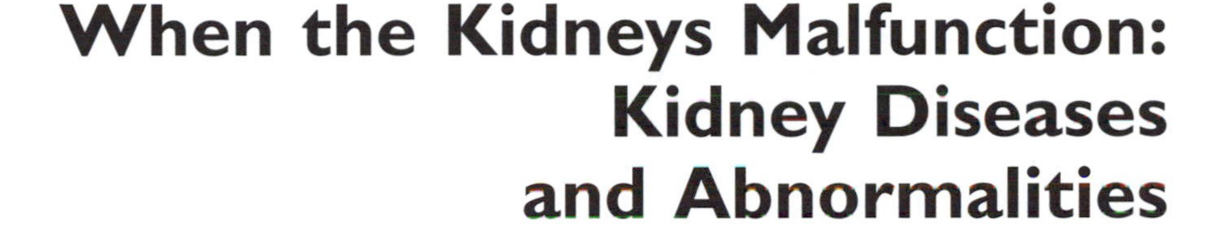

When the Kidneys Malfunction: Kidney Diseases and Abnormalities

Chapter 1 discussed the crucial role of the kidneys in ridding the body of toxins and maintaining a healthy water and chemical balance. Keeping in mind the importance of kidney function, imagine what would happen if they fell prey to disease or, worse yet, stopped working altogether. As the kidneys ceased their filtration, toxic substances would begin to build up in the blood, threatening precious organs and tissues. Meanwhile, necessary nutrients would seep into the urine and be flushed away. Fluid levels in the body would drop or rise, resulting in dehydration or oversaturation.

Fortunately, the body has a built-in backup system: If one kidney is ravaged by disease, the other assumes its role and takes charge of filtration and homeostasis. But what if disease cripples both kidneys? The results can be catastrophic. As the kidneys shut down, toxic chemicals build up in the blood. The result is end-stage renal disease (ESRD), a potentially fatal disorder that requires immediate dialysis or a kidney transplant (more on ESRD, dialysis, and kidney transplants appears in Chapter 5).

Kidney disease is a serious health threat in the United States, ranking ninth among the leading causes of death between 1999 and 2000. More than 20 million Americans suffer from chronic kidney disease, according to the National Kidney Foundation (2003). No one is immune from kidney disease, but adults over the age of 50, African Americans, Native Americans, Hispanics, Asians, and Pacific Islanders are most susceptible.

WHAT CAUSES KIDNEY DISEASE?

A number of diseases and conditions can interfere with the kidneys' ability to remove wastes. Some diseases attack quickly (acute), while others slowly devastate the kidneys over a period of many years (chronic). A disease may focus its attack specifically on the kidneys (like kidney cancer or polycystic kidney disease) or cause kidney damage as a peripheral effect of its symptoms (for example, **gout** can deposit a dangerous buildup of uric acid in kidney tissues).

Undoubtedly the two greatest threats to kidney function are diabetes and high blood pressure, which together account for more than 60 percent of kidney disease cases each year, according to the American Society of Nephrology (2001). Diabetes and high blood pressure go hand in hand, and together they can destroy the tiny filtering units, or glomeruli, that rid the blood of dangerous toxins. The other 40 percent of kidney disease cases stem from a variety of conditions and disorders. Stones, cysts, and other growths block fluid movement and leave the kidneys more susceptible to infection. Cancerous cells can destroy kidney tissue, just as they attack other organs of the body. Diseases that damage the glomeruli leave the filters scarred or inflamed to the point where they are unable to adequately clean the blood. And certain medications, when used over a long period of time, can damage or even destroy the kidneys.

Some people are born with abnormalities: misshapen, missing, or poorly functioning kidneys that can cause a lifetime of problems. Others inherit kidney diseases from their parents, and run the risk of passing these diseases on to their own children. The majority of people with kidney problems are born with healthy, functioning organs, but develop a kidney-damaging infection or disease at some point in their lives.

WHAT ARE THE WARNING SIGNS OF KIDNEY DISEASE?

Every kidney-related disease has its own roster of symptoms, each of which is a function of how the disease acts upon the body. Some symptoms mimic those of other disorders—for example, swelling in the face, ankles, and wrists indicates fluid retention, a sign that the kidneys are not filtering out enough fluid into the urine. But swelling can also occur during pregnancy, or it can be caused by excessive salt in the diet. Likewise, painful urination is often a sign of kidney disease, but it also can herald a urinary tract infection or certain sexually transmitted diseases (STDs).

Urine is one of the most important indicators of kidney function, because doctors can quickly analyze its composition and detect whether the kidneys are excreting too much or too little of a particular substance. An examination of the urine can reveal white blood cells, a sign of infection; red blood cells (hematuria), which indicate bleeding somewhere in the urinary tract; or abnormal amounts of protein (proteinuria), a sign of glomerular leakage.

If bleeding originates in the bladder, kidneys, or elsewhere in the urinary tract, the urine may appear a red or ruddy brown color. This is called **gross hematuria**, meaning that blood exists in a large enough quantity to be seen with the naked eye. Although the sight of bloody urine can be frightening to a patient, looks can be deceiving. As little as 0.03 ounces (1 milliliter) of blood is enough to tint the urine red. Blood may also be present in such small quantities that it cannot be detected unless viewed under a microscope. This is called **microscopic hematuria**.

Is hematuria a cause for concern? That depends on the amount of blood and its cause (see "Possible Causes of Hematura"). Something as benign as vigorous exercise can shed trace amounts of red blood cells into the urine. Doctors generally do not get concerned unless they see three or more blood cells within a high-powered field of the microscope. A woman's menstrual period can also turn the urine bloody. And certain foods (such as beets), medications, and food dyes can tint the urine a reddish color. In most cases, hematuria goes away on its own within twenty-four hours. But when coupled with other symptoms (for example, urinary urgency, incontinence, or loss of appetite), blood in the urine may indicate a more serious problem. Hematuria can warn of a urinary tract infection, glomerular disease,

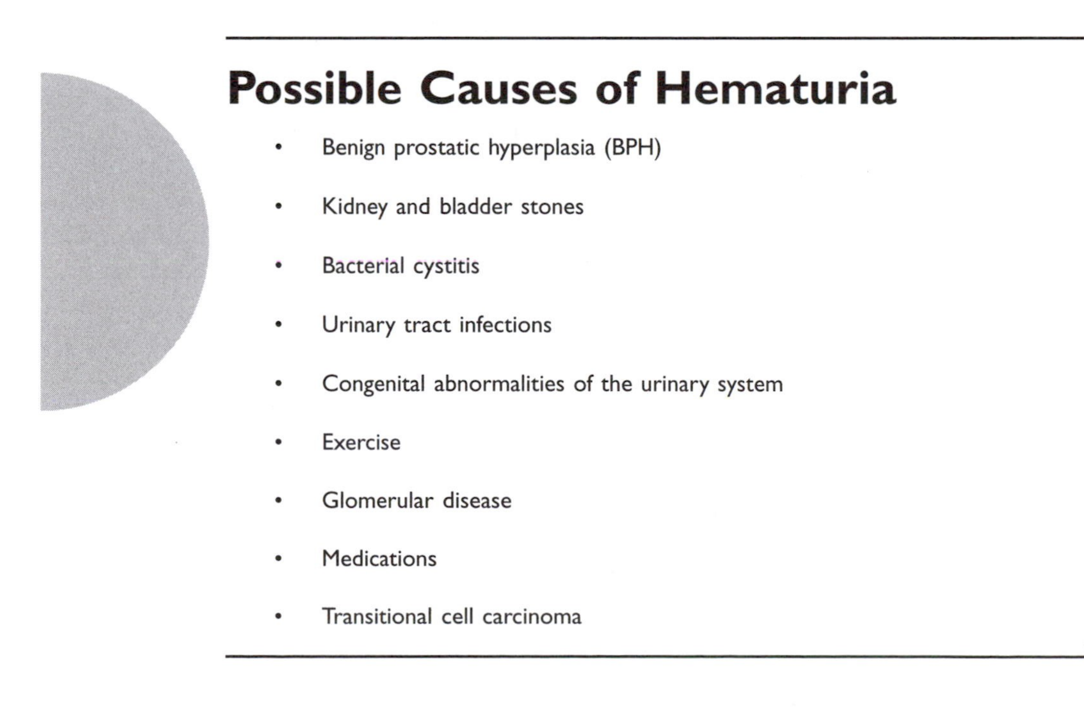

Possible Causes of Hematuria

- Benign prostatic hyperplasia (BPH)
- Kidney and bladder stones
- Bacterial cystitis
- Urinary tract infections
- Congenital abnormalities of the urinary system
- Exercise
- Glomerular disease
- Medications
- Transitional cell carcinoma

Possible Causes of Proteinuria

- Diabetes
- Hypertension
- Glomerulonephritis
- Amyloidosis
- Fever
- Dehydration
- Strenuous exercise
- Emotional stress

kidney stones, or the potentially deadly transitional cell carcinoma (a type of kidney or bladder cancer).

Proteinuria, or abnormally high levels of protein in the urine, also indicates a kidney problem (see "Possible Causes of Proteinuria"). Normally, most proteins are too big to pass through the glomeruli. But proteins can leak out through damaged filters and be mistakenly excreted in the urine. People who lose more than 150 mg of protein per day are said to suffer from proteinuria. Like hematuria, the causes of proteinuria can range from the completely benign (exercise or stress) to the potentially dangerous (kidney disease). The most common causes are high blood pressure (hypertension) and diabetes, but other diseases, like **amyloidosis** (abnormal protein deposits in the kidneys), can also be to blame. Benign cases usually resolve themselves, but kidney disease will eventually worsen if left untreated.

In patients who have proteinuria, the urine may appear foamy when it hits the water. This foaming occurs because protein alters the surface tension between urine and water. Patients may also experience swelling in their hands, feet, abdomen, and face, called edema. The swelling results from the protein albumin leaking into the urine. Albumin is one of the smaller proteins in the body and is thus one of the first to slip through when the glomeruli are even slightly damaged (called **microalbuminuria**). Albumin helps the blood absorb excess fluid from the body. When too much of the protein is excreted in the urine, the blood is unable to soak up enough fluid. That fluid tends to pool in the extremities and face, causing those areas to swell. Doctors can test for the presence of excess albumin and other proteins by analyzing a 24-hour urine sample under a microscope.

HOW IS KIDNEY DISEASE DIAGNOSED?

Physicians may refer patients who suffer from any of the symptoms discussed in the previous section to a **nephrologist**, a doctor who specializes in diseases of the kidneys. "Nephros" comes from the Greek word for kidney. "Renal," derived from *renes,* the Latin word for kidneys, is also a word often associated with kidney diseases and disorders. The nephrologist's first task is to take a complete medical history to determine whether a patient is at risk for hereditary kidney disease, and to find out whether he or she has been taking medications that are known to inhibit kidney function. The next step is a urine or blood test, which helps the doctor determine what type of kidney dysfunction is causing the symptoms. The same tests are used to diagnose other urinary system diseases, and will be referred to in later chapters.

Urine Tests

URINALYSIS

Usually the first line of investigation when any urinary disease is suspected, a urinalysis is an examination of the urine for the presence of certain chemicals or cells. As discussed in the last section, red blood cells (hematuria) indicate bleeding somewhere in the urinary tract, white blood cells signal an infection, and protein (proteinuria) is related to poor kidney function.

When a urine sample is collected in a small plastic cup, the doctor first screens it visually to make sure the color, odor, and concentration are normal (see Chapter 1 for a discussion on the characteristics of healthy urine). Then he or she analyzes the composition, looking for signs of excess protein, glucose, red and white blood cells, bacteria, or chemical crystals (which would indicate kidney stones, discussed in the section of this chapter on kidney stones). The doctor will also check the urine pH—a measure of acidity or alkalinity. In a healthy adult, the urine pH level ranges from 4.6 to 8.0, with a normal average of 6.0. A urine pH of 4 is highly acidic, 7 is neutral (neither acidic nor alkaline), and 9 is highly alkaline. Alkaline urine could signal, among other things, a urinary tract infection; acidic urine could be caused by uncontrolled diabetes, diarrhea, or dehydration. If the urinalysis reveals any sort of irregularity, the doctor will probably order one of these follow-up tests:

24-HOUR URINE COLLECTION

This test is similar to a normal urinalysis, except that instead of providing one sample, the patient collects all of his or her urine over a 24-hour period. A 24-hour urine collection allows the doctor to determine how much of a particular substance, for example protein, a patient is excreting during a given day. Doctors use a 24-hour sample to check for microalbuminuria, a sign of diabetic kidney disease, or to pinpoint what type of chemical (i.e., calcium, uric acid, or oxalate) is behind kidney stone formation.

CLEARANCE STUDY

Using a combination of urine and blood samples, with a **clearance study** the doctor is able to determine how efficiently a patient's kidneys are cleaning the blood. Specifically, clearance tests measure the kidney's glomerular filtration rate, or the amount of fluid filtered through each glomerulus per minute. The higher the glomerular filtration rate, the faster the kidneys are working to remove substances like protein, urea, and creatinine from the blood. The slower the rate, the slower the kidneys are removing these substances. A slow glomerular filtration rate may indicate kidney damage. The procedure for each clearance study varies, but typically consists of a 24-hour urine collection followed by a blood test.

Another type of clearance study, a urine osmolality test, measures the number of dissolved particles in a given volume of urine to determine how well the kidneys can produce a more concentrate or dilute urine. Referring back to the first chapter, the kidneys either add or remove fluid from the filtrate when the body has become dehydrated or overly saturated. The more dilute the urine, the more fluid the kidneys are filtering out. Likewise, the more concentrated the urine, the less fluid the kidneys are filtering out. An inability to regulate the body's fluids is yet another indication that the kidneys are not working properly. This type of test usually requires a "clean catch"; that is, urine collected in midstream. Disposing of the initial stream of urine prevents the sample from becoming contaminated with bacteria that are normally present in the urethra.

Blood Tests

Just as doctors use urine tests to determine whether the kidneys' filters are allowing necessary nutrients to escape into the urine, they use blood tests to look for substances (including creatinine and urea) in the bloodstream that healthy kidneys would normally filter into the urine.

BLOOD UREA NITROGEN (BUN) TEST

When proteins from food are broken down in the liver, ammonia is created as a byproduct. Because ammonia is too toxic for the body to process, the liver converts it into urea so that it can be excreted by the kidneys (see Chapter 1). Nearly all of the urea that filters through the kidneys should wind up excreted in the urine. If too much urea remains in the blood, as evidenced by high BUN results, the kidneys are most likely not filtering properly.

CREATININE TEST

Creatinine is produced as a byproduct of muscle metabolism. Like urea, it is normally filtered out of the blood by the kidneys and excreted in the urine. Because creatinine levels in the blood remain consistently stable, regardless of dietary changes, any elevation is a good indication of kidney damage.

ERYTHROPOIETIN (EPO) TEST

EPO is a hormone produced by specialized cells in the kidneys. When oxygen levels in the body drop, the kidneys respond by producing EPO, which in turn triggers stem cells in the bone marrow to produce red blood cells. As the number of red blood cells in the body rises, the blood is better able to carry oxygen. Low levels of EPO on this test indicate poorly functioning kidneys.

These are the three most common blood tests, but doctors can also determine how well the kidneys are functioning by checking levels of sodium, potassium, chloride, calcium, magnesium, protein, uric acid, or glucose in the blood (see Table 4.1).

Cystoscopy

If any of the urine or blood tests show higher or lower than normal levels of one or more substances, the doctor may recommend further tests. Cystoscopy relates more to the bladder and urethra than to the kidney, but it may be used to rule out bladder disorders when the cause of the problem is unknown. In this procedure, a narrow tube called a cystoscope is inserted through the urethra into the bladder. A tiny camera on the end of the cystoscope allows the doctor to examine the inside of the bladder and urethra for tumors, stones, and ulcers, as well as to check bladder capacity. If a tumor or stone is spotted, the doctor can immediately perform a biopsy or remove the growth with a special knife attached to the cystoscope.

Radiology

Just as doctors use x-rays to view the inside of the body when looking for broken bones, they use the same technology to look for stones and other masses inside the kidneys, ureters, and bladder, and to check that these organs are the right size and shape.

Intravenous Pyelogram (IVP) or *Intravenous Urography (IVU)*. An IVP or IVU (see photo) is the most widely used imaging technique for visualizing the urinary system. In this procedure, an iodine-containing contrast dye injected into a blood vessel allows the doctor to view an image of the patient's entire urinary tract on x-ray film. A series of x-rays reveals how fluid moves through the kidneys, ureters, and bladder. The dye displays the size and position of

TABLE 4.1. Blood Tests and What They Mean

Test	Normal Range (in Adults)	Possible Causes of Abnormal Results
Blood Urea Nitrogen (BUN)	7 to 20 mg/dl (milligrams per deciliter)	Kidney disease (e.g., glomulonephritis) Kidney failure Excess or too little protein in diet Urinary tract obstruction (i.e., stones or tumor) Malnutrition
Creatinine	0.7 to 1.5 mg/dl (milligrams per deciliter)	Dehydration Diabetic nephropathy Glomerulonephritis Reduced blood flow to kidney (i.e., from shock) Kidney failure Urinary tract obstruction
Erythropoietin (EPO) Test	0 to 19 mU/ml (milliunits per milliliter)	Low levels may mean chronic kidney failure
Sodium Test	136 to 145 meq/l (milliequivalents per liter)	Dehydration Diabetes Congestive heart failure Diarrhea or vomiting Kidney disease
Potassium Test	3.5 to 5 meq/l (milliequivalents per liter)	Crush injury (tissue trauma) Kidney failure Metabolic or respiratory acidosis Dietary deficiency of potassium Diarrhea or vomiting Renal tubular acidosis Diuretic use
Chloride Test	96 to 106 meq/l (milliequivalents per liter)	Dehydration Eclampsia Renal tubular acidosis Burns Congestive heart failure Excessive sweating Overhydration Vomiting
Calcium Test	8.5 to 10.3 mg/dl (milligrams per deciliter)	Hyperparathyroidism Hypoparathyroidism Vitamin D intoxication or deficiency Rickets Kidney failure

TABLE 4.1 *(continued)*

Test	Normal Range (in Adults)	Possible Causes of Abnormal Results
Magnesium Test	1.8 to 3.0 mg/dl (milligrams per deciliter)	Addison's disease Chronic kidney failure Dehydration Alcoholism Chronic diarrhea Hemodialysis
Protein Test	6 to 8 g/dl (grams per deciliter)	Malnutrition Nephrotic syndrome Chronic inflammatory disease (i.e., rheumatoid arthritis) Certain cancers
Uric Acid Test	Men: 3–9 mg/dl Women: 2.5–7.5 mg/dl (milligrams per deciliter)	Acidosis Alcoholism Diabetes Gout Hypoparathyroidism Kidney failure Too much exercise
Glucose Test	65 to 110 mg/dl (milligrams per deciliter)	Diabetes mellitus Hyperthyroidism Hypothyroidism Excessive or insufficient food intake

Note: These values vary by method.

the bladder and kidneys; reveals any abnormal masses, such as stones or tumors; and looks for any damage to the urinary tract caused by injury.

Computed Tomography (CT) Scan. For patients who are sensitive to the iodine-containing dyes used in an IVP, a CT scan is another effective method for detecting kidney stones, infections, and other abnormalities. This procedure uses a series of x-ray beams to create a 3-D computerized image of the kidneys (see photo in color insert). The CT is so detailed that it can display an organ from all sides on a video screen.

Ultrasound. During a woman's pregnancy, her obstetrician uses high-frequency sound waves, called ultrasound, to view an image of her growing baby on a video screen. That same technology is used to determine the size and shape of the kidneys, ureters, and bladder, and to look for cysts and other masses in the kidneys.

Magnetic Resonance Imaging (MRI). Huge, hollow MRI machines look almost like tunnels into which a patient slides on his or her back. The tunnel is surrounded by enormous magnets, which produce a powerful magnetic field. A com-

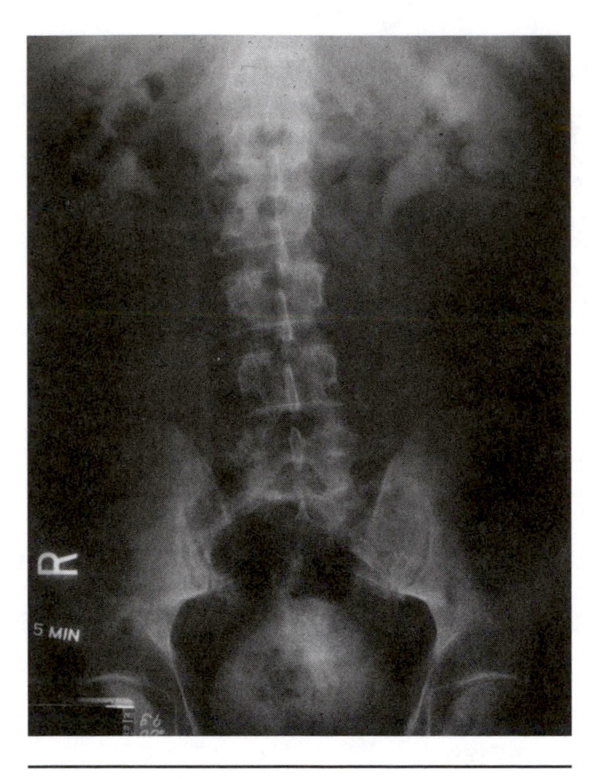

Intravenous pyelogram showing ureters, kidneys, bladder, and stones. © Custom Medical Stock Photo.

bination of this magnetic field and radio waves enables doctors to look through bones and view cross-sectional images of organs and tissues. The MRI can easily distinguish tumors from healthy tissues and determine how well blood is flowing to the organs.

KIDNEY ABNORMALITIES

Sometimes, people are born with missing, misshapen, or malfunctioning kidneys. The problem may be hereditary, caused by a genetic defect passed down from one or both parents, or it may originate with the patient. Abnormalities usually emerge in the developing fetus. In the first four weeks of a fetus's life, tiny stems within its pelvis grow and form the shape of the kidneys. Eventually, the kidneys move upward until they reach their final home in the flank area just beneath the ribcage (for more on urinary system development in the fetus, see Chapter 2). But a number of defects can affect not only how the kidneys grow but also where they are located.

Ectopic kidneys develop in the wrong part of the body. Rather than moving into the flank area, they may get stuck in the pelvis or rise too far and wind up in the chest cavity. This condition can be extremely painful, especially if the kidneys begin to crowd the lungs or other organs. Sometimes, both kidneys wind up on the same side of the body. If the kidneys are located in the wrong spot, they will most likely be unable to drain urine into the bladder. Fortunately, doctors can usually correct the problem with surgery.

Another abnormality occurs when, instead of one stem, two stems form on one side of the fetus and merge together into one double kidney. The double kidney may be too large to function, or one of the kidney segments may be nothing more than a nonfunctioning cyst, in which case surgery may be necessary to avoid infection and preserve kidney function.

In the case of a horseshoe kidney (see photo in color insert) the kidneys fuse together at the base as they rise from the pelvis, forming a U or horseshoe shape. People with this congenital abnormality often wind up with painful kidney stones, infections, blood in the urine, or poor urine drainage. Occasionally, surgery is required to correct the problem.

In a disorder called polycystic kidney disease (PKD), cysts, or fluid-filled sacs, form in the kidneys. PKD is passed via a defective gene from parent to child. The recessive form of this disease often originates in the womb and may affect both kidneys, in which case the baby is unlikely to live much past birth. In the more common dominant form of the disease, symptoms don't usually develop until later in life, and the prognosis for long-term survival is much more optimistic. More information on PKD is provided later in this chapter.

In some instances, only one of the stems develops fully and a baby is born with only one kidney (see Chapter 2 for more on kidney development in utero). When this occurs, the single kidney usually takes over for its missing twin. But in the rare cases when a baby is born without either kidney, death is inevitable.

Finally, an anomaly may occur not within the kidney itself, but at the point where the kidney and ureter connect. Ureteropelvic junction obstruction, as it is called, blocks urine from draining properly out of the kidney through the ureters. Often, the affected kidney must be removed and a drainage tube implanted until the healthy kidney can take over.

DIABETIC KIDNEY DISEASE (NEPHROPATHY)

Diabetes (also known as diabetes mellitus) occurs when the pancreas is unable to manufacture enough of the hormone insulin. Without insulin, the body can't properly process sugar for energy. Excess sugar, called glucose, remains in abnormally high amounts in the blood. Diabetics are more than twice as likely as people without diabetes to develop high blood pressure, according to the American Diabetes Association (2001). High blood pressure forces more blood through the glomeruli of the kidneys (called hyperfiltration), which slowly damages these filters. After years of overwork, the glomeruli start to show signs of wear, leaking albumin and other proteins into the urine. Eventually, the kidneys are unable to filter wastes from the bloodstream. In a vicious spiral, as the kidneys become more and more damaged, they in turn raise blood pressure even further by weakening the capillaries of the glomeruli. Ultimately, the kidneys shut down completely, and the patient must either undergo regular dialysis or a kidney transplant (see Chapter 5).

There are two types of diabetes: type 1 (insulin-dependant) and the more common type 2 (noninsulin-dependant). Both types of diabetes can damage the kidneys, but type 1 tends to carry a higher risk because the disease emerges earlier in life and therefore exacts a greater toll on the kidneys.

Symptoms and Diagnosis of Diabetic Kidney Disease

Patients may experience no symptoms in the early stages of diabetic kidney disease, as the kidneys compensate for glomerular damage by simply working harder. In fact, symptoms may not appear until the kidneys are close to failing. The only way to detect kidney disease in many cases is by conducting annual blood and urine screenings. When the kidneys are in trouble, examination of the blood will reveal high blood urea nitrogen (BUN) and creatinine levels, because the kidneys are unable to filter out these wastes. The urine will most likely contain excess protein (proteinuria) that has leaked from the kidney's filters. Other symptoms may include high blood pressure; swelling in the hands, feet, or legs (from excess fluid in the body); an increased urge to urinate; nausea and vomiting; weakness or anemia; loss of sleep; or poor appetite.

Treatment

No cure exists for diabetic kidney disease, but several treatments can slow its progression to kidney failure. Right now, the recommended regimen focuses on controlling blood pressure and regulating blood glucose levels. According to the National Kidney Foundation, high blood pressure may be the most important key in determining which diabetics develop kidney disease. To keep blood pressure at a healthy level (under 130/85), doctors often recommend a class of drugs called angiotensin converting enzyme inhibitors (ACE inhibitors), which not only lower blood pressure but also reduce proteinuria, thereby slowing degeneration of the glomeruli. Scientists have found ACE inhibitors beneficial even to patients who don't have high blood pressure, because the drug appears to directly protect the glomeruli from deterioration. In addition to medication, patients can control blood pressure by eating a healthy diet, losing weight, exercising, and reducing the amount of salt in their diet.

Controlling blood glucose levels can also delay kidney damage, especially when started early. In the Diabetes Control and Complications Trial (DCCT), a study of 1,400 diabetics conducted between 1983 and 1993, patients who maintained close-to-normal blood glucose levels (between 70 and 180 mg/dl, depending on food intake) slowed the progression of their kidney disease by 50 percent. Diabetics can control their blood glucose levels by administering insulin shots as needed, dieting and exercising regularly, and keeping up with their doctor visits. With regard to diet, diabetics may want to avoid high-protein foods like red meat, milk, and cheese, which force the kidneys to work harder.

If the treatments discussed here are unsuccessful and the kidneys fail, the last resort is either dialysis or a kidney transplant. Dialysis works well in the short term, but, ultimately, a transplant is the only way to restore kidney function. Back in the 1970s, doctors were reluctant to perform kidney transplants on diabetics, because they felt that damage caused by the disease would negate the benefits of transplantation. Today, as patients are able to better control their diabetes and live longer, fuller lives, transplantation is a viable option, and success rates among diabetics are equal to those in the general population.

Looking to the future, scientists are trying to determine who is susceptible to diabetic kidney disease in order to prevent it. To that end, they are analyzing cells from diabetic patients to gain an understanding of why one person develops kidney disease and another does not. Until scientists discover the key to diabetic kidney disease, the best preventive measure for diabetics is to control their blood pressure and glucose levels early in the course of the disease.

KIDNEY STONES

Kidney stones may be small in size, but they are responsible for some pretty startling statistics: Over 1 million new cases reported in the United States each year; more than 200,000 hospitalizations each year, according to the American Foundation for Urologic Disease (2001); and an estimated 10 percent of the American population affected during the course of their lives.

The incidence of kidney stones has risen over the past twenty years. Doctors aren't sure why, but it may have to do with the evolution of our diets. Fatty, protein-rich fast foods have become a staple of the American diet over the last couple of decades, and protein, as we'll discuss later in this chapter, is one of the prime culprits in kidney stone formation.

How Kidney Stones Form

Stones (also called **calculi**) are formed when chemical crystals build up in the urine, separate, and move into the kidneys or other organs. These crystals are more likely to build up when the urine level is low, as is the case when the body lacks sufficient liquid volume. The kidneys conserve fluid by excreting a smaller volume of more highly concentrated urine. Think of pouring sugar into a glass of iced tea. When a small amount is added, it dissolves easily. But as more is added, the crystals start to settle to the bottom. The less liquid in the glass, the more quickly the crystals will separate from the tea, which is why urine concentration is so important to stone formation.

Whether or not an individual will develop stones also has to do with an absence or low concentration of special substances in the urine, such as **citrate** (a salt or ester of citric acid that binds calcium ions) and magnesium, which normally block crystal formation. If the body does not produce enough of these crystal-inhibiting chemicals, or if the urine is oversaturated with crystals, the conditions are ripe for stone formation.

Some crystals are so small that they move through the body without ever being detected. Others cluster together, forming a mass with a small stone inside. As new crystals continue to attach themselves to the outside of this mass (picture individual snowflakes falling and catching on to one another), the ball will grow larger and larger. Usually, the growing crystal formation starts out attached to the wall of the kidney's filtering tubule, but as it gets larger, the stone breaks off from the kidney and passes into the ureter leading to the bladder. The stone will continue to move until it reaches a narrow spot and becomes stuck. A stone moving through the urinary system might feel like a large pebble stuck in a person's shoe. As the individual walks, the stone rubs against the skin, creating a painful friction. A kidney stone is even more painful, because it rubs against sensitive tissue on the inside of the body.

Chemicals That Cause Kidney Stones

CALCIUM STONES

Normally, most of the calcium filtered through the glomeruli of the kidneys is reabsorbed into the bloodstream. When the body absorbs too much calcium, it excretes abnormally high amounts in the urine, a condition doctors refer to as **hypercalciuria**. Foods, medications, and metabolic disorders can all increase calcium levels in the urine. These include:

- Overactive **parathyroid glands** (hyperparathyroidism), a malfunction of the glands that regulate how much calcium is absorbed by the intestine.
- Bone cancer forces calcium out of the bones and into the bloodstream.
- Excess vitamin D causes the body to absorb too much calcium from foods.
- The use of steroids and diuretics elevates calcium concentrations in the urine.

Calcium-based stones are most often composed of a combination of calcium and oxalate, phosphate, or uric acid. About 75 percent of calcium stones are made up of calcium combined with the metabolic waste product, oxalate. Between 10 and 15 percent of the oxalate concentration in the urine originates in the diet, contained in foods like rhubarb, tea, nuts, chocolate, and beans. Alternately, calcium oxalate stones may result from an excess of oxalate in the urine, called **hyperoxaluria**. A smaller percentage of calcium stones are composed of a combination of calcium and either phosphate or uric acid.

URIC ACID STONES

When the body digests proteins, it produces a nitrogen byproduct called purine, out of which uric acid crystals form. High concentrations of these crystals, or a condition known as **hyperuricuria**, can result in the formation of uric acid stones. So it makes sense that people who eat a lot of meat, which is high in protein, are more likely to suffer from uric acid stones. Patients with uric acid stones usually have very acidic urine, which can precipitate stone formation.

A number of conditions and medications may also lead to hyperuricuria:

- Gout, a painful arthritic condition in which excess uric acid deposits in the joints or kidneys
- Diuretics and chemotherapy drugs
- Large quantities of alcoholic beverages
- Fasting
- Chronic diarrhea
- Leukemia or other blood diseases

CYSTINE STONES

According to the American Foundation for Urologic Disease (2001), about 1 in 20,000 people suffers from an inherited metabolic disorder called **cystinuria**. This condition causes an abnormal transport in the kidneys and intestines of the amino acids cystine, ornithine, lysine, and arginine. The kidneys end up excreting an abnormal amount of cystine into the urine. Whereas a healthy adult may excrete about 100 milligrams of cystine per day, a patient suffering from cystinuria will urinate as much as 400 to 500 milligrams per day. Cystine doesn't dissolve well in urine to begin with, and an excess of the chemical only increases the likelihood that stones will form.

STRUVITE STONES

Struvite stones almost always occur as the result of a urinary tract infection (see Chapter 9 for more on urinary tract infections), and because women are more prone to infection, they are also more likely than men to develop this type of stone. When bacteria are present in the urinary system, they secrete certain enzymes that increase the amount of ammonia and magnesium in urine. The excess ammonia is what forms struvite stone crystals. When a stone grows large enough, it can block kidney drainage, creating a breeding ground for even more bacteria and a favorable condition in which more stones are likely to develop.

Symptoms of Kidney Stones

An early-stage kidney stone may be too small to produce pain or other symptoms. As the stone grows larger, the first sign is usually a colicky pain in the back and side or lower abdomen that lasts from five to fifteen minutes at a time. Eventually, the stone is big enough to break free from the kidney, and it slowly makes its way down the ureter toward the bladder. The friction ignited by the movement of stone against sensitive tissue causes those tissues to become inflamed and swollen. The pain grows along with the stone, and anyone who has suffered a kidney stone will attest that large stones can be excruciating.

Pain is the most obvious symptom of a kidney stone, but other signs may include blood in the urine, as the stone irritates the lining and tubes of the kidney; more frequent urination or a burning sensation during urination as the stone moves through the ureter; and fever and chills if the stone blocks the flow of urine and the kidney becomes infected.

Diagnosing Kidney Stones

Doctors have been able to identify and treat kidney stones for centuries, but recent medical developments have placed a wealth of new diagnostic and treatment tools at their disposal. A patient with a kidney stone will usually go to the doctor complaining of back pain or blood in the urine. To determine the source of the problem, the doctor will begin by asking whether the patient has a personal or family history of kidney stones. Next, a physical examination will rule out other conditions that have similar symptoms (for example, appendicitis). Then, the doctor will take a blood test to check urea nitrogen, creatinine, electrolyte, and uric acid levels. The urine will be analyzed for bacteria (which would indicate an infection), pH levels (patients with uric acid stones usually have acidic urine, whereas patients with struvite stones usually have alkaline urine), and the presence of stone-forming chemicals.

Before recommending any treatment regimen, the doctor must pinpoint the size, type, and location of the stone. Intravenous pyelogram (IVP) remains the preferred method for hunting down stones, measuring their size, and determining to what degree they are obstructing urine flow. An ultrasound can also locate a stone within the kidney or ureter, but it is somewhat limited in determining the degree of obstruction. Increasingly, doctors are turning to a new method, called the helical computerized tomography (helical CT), for the initial assessment of kidney stones. The helical CT, which uses a series of x-ray beams to produce two-dimensional images of the urinary system, is not only faster and more sensitive than IVP, but it also can locate stones anywhere in the urinary tract.

Treating Kidney Stones

Most kidney stones are so small that they pass through the body without treatment and in some cases go unnoticed. Doctors often recommend that patients with small stones drink plenty of water (2 to 3 quarts per day), to flush out stone-forming chemical crystals. If a stone is too large to pass on its own (larger than 5 mm in diameter), or if it blocks the flow of urine, causes a urinary tract infection, or damages the kidneys, it must be removed via a surgical procedure.

At one time, open surgery was the only option for removing stones. Stone surgery was painful and required a lengthy recovery period. But over the past fifteen years, the less invasive methods of extracorporeal shock wave lithotripsy and endoscopic surgery have replaced open surgery in virtually all cases.

EXTRACORPOREAL SHOCK WAVE LITHOTRIPSY

From the Greek word for "stone crushing," extracorporeal shock wave lithotripsy (ESWL) literally crushes stones with a powerful electrical shock wave (18,000 to 24,000 volts). ESWL has been used in the United States since 1984, and it is the method preferred by the American Urological Association for removing small stones. A machine called a lithotripter focuses shock waves on the area in which the stone is located (see photo in color insert). The shock waves travel through the skin and body tissues and disintegrate the stone into tiny sand-like particles, which are easily passed from the body.

Two types of ESWL are commonly used to destroy stones. In one method, the patient lies in a bath of water while the shock waves are transmitted. In the other, the shock waves pass through a soft cushion on which the patient is lying. Doctors use x-rays or ultrasound to pinpoint the stone's location before beginning the ESWL procedure. The patient is awake, but usually receives anesthesia because the procedures can be uncomfortable. After treatment, doctors may insert a small tube, called a stent, through the patient's bladder and into the ureter to help the stone pass. Recovery from ESWL is relatively quick, and the side effects are minor. Some patients have blood in their urine for a few days after treatment and bruises on their back or abdomen where the shock waves were focused.

URETEROSCOPIC STONE REMOVAL

Stone removal through telescopes called ureteroscopes was first performed in the early 1980s, but use of these devices was somewhat limited at the time. Because the scopes were large and inflexible, they could not reach up the ureter and therefore could only remove stones that had moved to the junction between the ureter and bladder. But with the advent of flexible scopes not much bigger than the size of a needle, surgeons can now remove stones virtually anywhere in the urinary tract. The new ureteroscopes are threaded through the urethra and bladder until they reach the stone, and the surgeon removes the stone either with a cage-like device or by shattering it into tiny fragments with a laser attached to the scope.

PERCUTANEOUS NEPHROLITHOTOMY

If the stone is too large, or if stones are located in more than one place, doctors may try a surgery called percutaneous nephrolithotomy. In this procedure, the surgeon makes a tiny incision in the patient's back and tunnels through to the kidney. The stone is removed using a tool called a nephroscope. In cases where the stone is especially large, an ultrasonic or electrohydraulic energy probe may be needed to break it into smaller pieces for removal. Because this procedure is more invasive than other stone-removal methods, it requires a hospital stay of several days.

MEDICATION

Doctors may prescribe narcotics such as codeine, morphine, and meperidine (Demerol), or anti-inflammatory agents like aspirin or ibuprofen, to help patients deal with the pain of passing a kidney stone. In some cases, medications can also inhibit stone formation. What medication the doctor prescribes depends on the type of stone:

Calcium Stones. Diuretics, which treat high blood pressure by removing excess fluid from the body, can also reduce the amount of calcium excreted by the kidneys into the urine. Doctors also prescribe potassium citrate for calcium stones because it reduces calcium excretion and inhibits crystal formation, and cellulose phosphate, which binds to calcium in the intestines and prevents it from leaking into the urine.

Uric Acid Stones. When stones are caused by overly acidic urine, patients can take sodium bicarbonate or potassium citrate to reduce acidity. Doctors also prescribe the drug Allopurinol to reduce uric acid levels.

Struvite Stones. The goal in treating struvite stones is to eliminate the underlying infection with antibiotics. Once the stones have been removed, a drug called acetohydroxamic acid (Lithostat), along with an extended course of antibiotics, helps prevent the infection from recurring.

Cystine Stones. Cystine stones are the most difficult stones to treat. Potassium citrate or sodium bicarbonate are somewhat effective, because they make the urine more alkaline and increase the solubility of cystine. If these drugs don't work, penicillamine or tiopronin can help the body more effectively dissolve cystine. Cystine stone sufferers are told to drink up to 12 glasses of water every day, even more so than patients with other types of stones, so the chemical will dissolve more effectively in the urine.

Preventing Kidney Stones

About half of the patients who suffer from kidney stones will have a recurrence within five to ten years after treatment, according to the American Foundation for Urologic Disease (2001). The good news is that there are ways to prevent future episodes. The first step in prevention is to understand what chemical is causing the stones. Often, doctors will ask patients who are passing stones to hold a strainer under the flow of urine to collect the stones for analysis. Once the chemical component is isolated, preventive medication or lifestyle changes may ward off future episodes. One of the most effective ways to prevent stone formation, say doctors, is to drink plenty of fluids—at least 10–12 glasses a day—half of which should be water. In hot weather, patients are advised to add a couple of glasses per day to compensate for fluid lost through perspiration. Researchers say lemonade may inhibit stone formation because it contains citrate, but grapefruit and apple juices may not be wise choices because studies have found that they actually increase the risk of stone formation.

Diet, too, can play an important role in stone prevention. People who have uric acid stones may need to eat less meat, fish, and poultry, all of which can make the urine more acidic. Those who are prone to forming calcium oxalate stones may want to avoid foods high in oxalates, including organ meats (i.e., liver and kidneys), fish, beets, chocolate, nuts, rhubarb, spinach, strawberries, tea, and coffee. Following this rationale, it would appear to make sense to restrict dietary calcium in patients who have calcium stones, but actually, research has proven otherwise. Dairy and other calcium-rich foods don't appear to affect stone formation. Proteins and salt are considered risk factors, however.

Future Directions in Kidney Stone Research

Scientists are trying to predict who is at risk for kidney stones so that preventive efforts may be initiated earlier. To accomplish this, they must discover the role of genetics, if any, in stone formation. They must also discover new substances that block stone formation. For example, a peptide known as the crystal matrix protein has been found to inhibit calcium oxalate crystal formation in rats, but further research is needed to determine whether it has the same effect on humans.

KIDNEY CANCER

Cells constantly grow and divide according to set patterns. But when cancerous cells enter the picture, the orderly process of cell division turns chaotic, and cells are reproduced even when the body doesn't need them. These extra cells form a mass of extra tissue, called a tumor. Tumors are said to be malignant, or cancerous, when they have the ability to spread (called metastasis) throughout the body. Once cancer cells metastasize, they can run rampant through the body, destroying healthy tissue and eventually killing the patient.

Just as cancer strikes the lungs, throat, liver, bladder, and other organs, it can also attack the kidneys. According to the National Cancer Institute (2000), kidney cancer occurs most often between the ages of 50 and 70, and affects nearly twice as many men as women. Doctors don't know exactly what causes kidney cancer, but there are a number of suspected risk factors:

Smoking. Smokers are twice as likely to develop kidney cancer than non-smokers. The longer a person smokes, the greater his or her risk of developing kidney cancer.

Occupational Exposure. There is evidence that asbestos, which has been linked to lung cancer, also plays a role in the development of kidney cancer. Chemical components used in steel plants have also been shown to produce above-average rates of kidney cancer.

Obesity and High Blood Pressure. Obesity and high blood pressure have both been associated with kidney cancer, especially in men. Scientists are still unclear as to why there is a connection.

Radiation. Women who have undergone radiation therapy for uterine cancer may have a slightly higher risk of developing kidney cancer. Also, people who were exposed to thorium dioxide, a radioactive substance used in x-rays during the 1920s, have increased rates of the disease.

Dialysis. Patients who have been on dialysis for a long period of time may develop kidney cysts, one cause of renal cell cancer.

Von Hippel-Lindau (VHL) Disease. This inherited genetic defect increases the risk of developing renal cell cancer as well as other types of cancers.

Types of Kidney Cancer

Renal Cell Carcinoma. About 85 percent of kidney cancers are renal cell carcinoma, making it the most prevalent type of cancer affecting the kidney. Renal cell carcinoma develops in the cells that line the proximal renal tubules in the nephrons—the cells responsible for filtration and urine formation (see Figure 1.3 in Chapter 1). This type of cancer usually develops in only one kidney, but it can spread quickly to nearby blood vessels and invade other organs.

Transitional Cell Carcinoma. Transitional cell carcinoma is much more rare than renal cell carcinoma, affecting only about 10 percent of kidney cancer patients, but it can be a much more aggressive cancer. Malignant cells form in the renal pelvis—the part of the kidney that collects urine and sends it to the bladder—and/or the ureters. Transitional cell carcinoma is so similar to bladder cancer that treatments for the two diseases are often identical (see Chapter 6 for a discussion on bladder cancer).

Wilms' Tumor. Children under the age of 5 are most likely to be affected by Wilms' tumor, the most common type of childhood kidney cancer. This cancer may be found only within the kidney; in later stages, it may spread to the lymph nodes or other organs.

Other cancers may spread from the lung, liver, or another organ to the kidney. These are called secondary kidney tumors. Likewise, kidney cancer can spread to other organs, but the disease still carries the name kidney cancer, regardless of where the cancerous cells are located.

Symptoms of Kidney Cancer

Early detection can be difficult, because the disease often presents with no symptoms initially. Symptoms don't usually appear until the tumor has grown fairly large and has spread to other parts of the body. Once the cancer has begun to spread, the most common sign is blood in the urine. Other indicators may be a lump in the kidney area discovered during a physical exam, pain in the back just below the ribcage that does not go away, fatigue, loss of appetite, weight loss, fever, or anemia (low red blood cell count).

Diagnosing Kidney Cancer

Several decades ago, most kidney cancers were not diagnosed until patients visited their doctor in pain. But since the introduction of ultrasound and CT abdominal scans in the 1980s, the number of cases diagnosed incidentally during examinations for other conditions has risen dramatically.

Any time kidney cancer is suspected, the first step in the diagnostic procedure is to take a medical history and physical exam, followed by blood and urine tests. The doctor may also feel the patient's abdomen for lumps or other irregular masses. If any of these tests indicate the possibility of cancer, the doctor will probably look for masses in the kidneys, as well as in the ureters and bladder, via an ultrasound, CT scan, IVP, or MRI. A cystoscopy may be used to determine whether the cancer has spread. This procedure allows the doctor to view the interior of the urethra, bladder, and kidneys, and, if a tumor is located, to remove a small tissue sample (or biopsy) for evaluation.

Once a tissue sample is removed, a pathologist will examine it under a microscope to learn whether it is benign or malignant (cancerous). If the cells are malignant (see photo in color insert), the pathologist must determine their histology, or type. Knowing what types of cells make up the tumor, and how aggressively those cells will behave, allows the doctor to determine the patient's prognosis and to plan the most appropriate treatment course.

If the cancer is suspected to have spread, the doctor may also call for a chest x-ray, liver ultrasound, or bone scan. In addition, he or she may order a kidney function test to determine whether one kidney is strong enough to take over for the cancerous kidney, should it need to be removed.

Before treatment, the doctor will also need to know how far the tumor has progressed, which is called staging. Four stages are used to describe how kidney cancer progresses (the stages below apply to renal cell carcinoma specifically; other types of cancers use a slightly different staging order):

Stage 1. The cancer is very small (less than 2.75 inches, or 7 cm in diameter) and has not spread beyond the walls of the kidney.

Stage 2. The cancer has grown larger (greater than 2.75 inches, or 7 cm in diameter) but is still confined within the kidney.

Stage 3. The cancer has spread to nearby blood vessels, the adrenal gland, or lymph nodes around the kidney, but has not reached other organs.

Stage 4. The cancer has spread to nearby organs (for example, the bowel or pancreas), lymph nodes, or organs farther away (e.g., the lungs). Once the cancer has spread, the prognosis is usually not good.

Most patients diagnosed with kidney cancer are referred to a cancer specialist, or oncologist, for treatment. The course of treatment will depend on the size of the tumor and whether the cancer has spread.

Treatments for Renal Cell Carcinoma

SURGERY

Surgery is the most common, and most effective, way to treat renal cell carcinoma. More than 75 percent of patients who undergo surgery survive their cancer (Russo, 2000). Unfortunately, surgery is of little use when the cancer has spread to other organs. Three-quarters of patients with metastatic kidney cancer die within a year of diagnosis.

Nephrectomy is the name for surgery to remove the cancerous kidney. In a simple nephrectomy, the surgeon removes just the diseased kidney; in a radical nephrectomy, the surrounding adrenal gland and lymph nodes are removed along with the kidney; and in a partial nephrectomy, the surgeon removes just the diseased part of the kidney, sparing at least some of the organ's function. Radical nephrectomy was once considered the preferred method for kidney cancer surgery because it can quickly eradicate the cancer, even in cases where the tumor has already grown quite large. But, increasingly, doctors are turning to partial nephrectomy (otherwise known as nephron-sparing surgery), because it preserves kidney function without compromising the patient's life. When researchers at the Mayo Clinic in Rochester, Minnesota, compared outcomes for 328 patients, half of whom had had a radical nephrectomy and the other half of whom had had a partial nephrectomy, survival rates and postsurgical complications were nearly identical among the two groups. Furthermore, patients who underwent a partial nephrectomy were less likely to develop chronic kidney failure than those who had radical nephrectomies. Researchers still need to conduct additional follow-up with partial nephrectomy patients to determine whether cancer cells will recur in the portion of the kidney that remains.

Doctors may also opt for a less invasive form of kidney removal—laparascopic nephrectomy. In this procedure, a surgeon inserts a small telescope called a laparascope through a small incision and performs the operation using tiny instruments inserted through other small incisions. The doctor can view an image of the patient's kidney on a television screen to guide the operation. Laparascopic nephrectomy reduces the hospital stay and recovery time, and is less invasive than traditional nephrectomy surgery, but the procedure may not be as widely available because it requires special surgical training.

Before a nephrectomy, doctors may use arterial embolization to make the tumor easier to remove. In this procedure, small pieces of a special gelatin sponge or other material are injected through a tube called a catheter to clog the main blood vessel to the kidney. Deprived of blood and oxygen, the tumor will begin to shrink. Doctors may also use this procedure after surgery to relieve the patient's pain if part of the tumor could not be removed.

Surgery may be followed up by adjuvant, or added, therapy involving radiation, chemotherapy, or immunotherapy to destroy any microscopic cancer cells left behind.

IMMUNOTHERAPY (BIOLOGICAL THERAPY)

Immunotherapy is one of the most promising treatments for kidney cancer, because it uses the body's own immune system to destroy cancer cells. Drugs called biological response modifiers, which are either made by the body or recreated in a laboratory, stimulate white blood cells in the immune system to fight off the cancer. Interleukin-2 (IL-2) and interferon-Alpha are the two drugs most commonly used.

The side effects of immunotherapy are usually mild and may include flu-like chills, fever, muscle aches, loss of appetite, nausea, vomiting, and diarrhea. These symptoms typically disappear after the treatment is complete.

RADIATION THERAPY

Radiation is used against a variety of cancers to kill malignant cells and shrink tumors. Rays can either be directed at the tumor from a machine (external radiation therapy) or inserted into the body via tubes (internal radiation therapy). Radiation can be used alone or as an adjuvant therapy to chemotherapy or surgery. Unfortunately, this therapy has not been effective in treating renal cell carcinoma, but it has been used to relieve pain in patients whose cancer has progressed to the inoperable stage.

Side effects of radiation are more severe than those of immunotherapy and may include nausea, vomiting, diarrhea, fatigue, and red or dry skin around the treated area. Patients are also more prone to infection, because radiation kills healthy cells along with cancerous cells and may destroy some of the white blood cells that protect the body from infection.

CHEMOTHERAPY

Chemotherapy uses drugs—in pill, liquid, or intravenous form—to target and kill cancer cells. Like radiation therapy, this treatment is of little use against renal cell carcinoma. Out of the fifty or so chemotherapy drugs currently in use, only about four to six have shown any effectiveness in treating kidney cancer, and their effects have been limited. Doctors are also using "cocktails" combining various chemotherapy drugs.

Side effects from chemotherapy vary based on the type of drug used, but many patients complain of anemia (from the loss of red blood cells), greater susceptibility to infection (from the loss of white blood cells), nausea and vomiting, hair loss (from the loss of hair follicle cells), and fatigue.

HORMONE THERAPY

This treatment is similar to chemotherapy, but hormones (usually progesterone) rather than drugs are used to control the growth of cancerous cells. Side effects with hormone therapy are usually mild. The patient may experience changes in appetite or weight, swelling, or fluid retention, all of which normally disappear after the treatment is finished.

Future Directions in Renal Cell Cancer Detection and Treatment

In the past, doctors could do little to stop renal cell cancer progression when patients did not respond to traditional therapies. But in recent years, several experimental treatments have shown great promise when all other treatments have failed:

STEM CELL TRANSPLANTS

Stem cells are the building blocks of every tissue and organ in the human body, and they have shown great promise in repairing tissue damaged by diseases such as Parkinson's and Alzheimer's. Researchers say these fledgling cells also have the potential to shrink renal cell carcinoma tumors. Stem cells are taken from tissue donated by the patient's brother or sister, and the patient is given immunosuppressive drugs (like cyclosporine) to stop his or her immune system from rejecting the foreign tissue. Study results so far have been encouraging, but doctors still need to find a way to combat the problem of graft versus host disease, a serious complication of the transplant procedure in which the donor's immune cells attack the patient's healthy tissues.

VACCINE THERAPY

Vaccines stimulate the immune system to fight off diseases such as measles, diptheria, or chicken pox. But unlike traditional vaccines, cancer vaccines don't prevent disease—they treat it. Doctors take a small sample of cancer cells from the patient and fuse them, using an electric current, with white blood cells from a healthy donor. The vaccine is then injected into the patient. When the body spots the donor's cells, it recognizes them as foreign invaders and attacks. Because the donor cells are fused to tumor cells, the immune system learns to regard the tumor as a foreign invader as well. In this way, the vaccine teaches the patient's immune system to seek out and destroy cancer cells wherever they may hide in the body. In clinical trials, cancer vaccines have been successful in shrinking tumors, but the therapy is still under investigation.

GENE THERAPY

Like vaccine therapy, gene therapy takes cells from the patient, alters those cells in the lab, and returns them to the patient's body, where they train the immune system to hunt down and kill cancer cells. But in gene therapy, scientists add a gene to the patient's T cells (a type of white blood cell involved in the body's immune system response), which tricks the body into attacking any cancer cells they find as if they were foreign tissue. Genetically altered cells have been effective so far in laboratory studies, but clinical trials over the next several years will be needed to determine whether this therapy is safe and effective over the long term.

ANTIANGIOGENIC THERAPIES

These therapies, under investigation to combat several types of cancer, kill tumors by cutting off their blood supply. Among the most promising in kidney cancer research is thalidomide, the sleep aid banned in Europe in the 1960s because it caused birth defects. Trials at Memorial Sloan-Kettering Cancer Center in New York and Baylor University in Texas have found that thalidomide slowed progression of kidney cancer. Researchers are now trying out the drug at different stages of cancer development as well as in combination with chemotherapy.

TUMOR MARKERS

In addition to looking for new treatment options, scientists are trying to learn how to detect kidney cancer earlier, before it has spread. One method that is showing great promise is the use of tumor markers—substances present in abnormal amounts in the blood, urine, or tissues of cancer patients. Several markers are under investigation, but none has been isolated for detecting kidney cancer as of yet.

Treatments for Transitional Cell Carcinoma

Surgery is the most common treatment for transitional cell carcinoma. Depending on the size and location of the tumor, surgery can range from a minimally invasive laser therapy to complete removal of the kidney, ureter, and part of the bladder. Some of the most common surgical techniques are:

Segmental Resection. If the cancer has not spread, surgery to remove part of the ureter and kidney is usually successful.

Electrosurgery. The doctor uses an electrical current to burn the tumor and the area around it, then cuts out and removes the tumor with a sharp tool.

Laser Therapy. An intense beam of light focused on the tumor shrinks cancerous cells (electrosurgery and laser therapy can only be used when cancers are located on the surface of the renal pelvis or ureter).

Nephroureterectomy. If the cancer is too large to respond to the above therapies, doctors must surgically remove the entire kidney, ureter, and the ring of tissue surrounding the opening of the ureter into the bladder.

Depending on the stage and size of the tumor, doctors may also use radiation, immunotherapy, or chemotherapy, either as adjuvant treatments preceding or following surgery, or on their own.

Regardless of the type of therapy used, patients with transitional cell carcinoma require substantial follow-up, because as many as 50 percent will go on to develop bladder cancer. For those patients whose cancer is only on the surface of the kidney and/or ureter, survival rates are excellent—70 to 80 percent. But if the tumor grows deep or spreads to other parts of the body, only about 10 percent of patients may survive.

Treatments for Wilms' Tumor

Cancer is always devastating when it strikes a child, but fortunately, Wilms' tumor is nearly always treatable. As is the case with other cancers, treatment depends on the stage of the cancer and tumor size, as well as the child's age and health. The younger a child is when diagnosed, the less risk the child has of experiencing a relapse. If the cancer is small and has not spread, doctors can usually remove just the affected portion of the kidney (partial nephrectomy). If the cancer has grown larger, doctors may opt to remove the entire kidney. And if the cancer has spread to tissues around the kidney or to the blood vessel or renal sinus, a radical nephrectomy may be the only option. Once the cancer has spread beyond the kidney area, treatment becomes much more complicated, involving a combination of surgery and either chemotherapy or radiation therapy.

KIDNEY CYSTS

Cysts are fluid-filled sacs that may appear virtually anywhere in the body, including the kidneys. Kidney cysts are quite common, especially in people over the age of 50. Cysts may occur singly or in groups. Simple cysts are usually harmless and require no treatment, although doctors may perform an ultrasound to make sure that the cysts are not irregularly shaped, which could indicate a malignancy. If the ultrasound is inconclusive, doctors may also remove fluid from the cyst and examine it under the microscope to look for cancerous cells.

Sometimes, cysts grow so large or so plentiful that they virtually take over the kidneys, inhibiting filtration and other kidney functions. Two conditions that cause this proliferation of cysts are known as multicystic kidney disease and polycystic kidney disease.

Multicystic Kidney Disease

In the tiny developing fetus, one tiny misstep can drastically alter the way organs are formed. In the case of multicystic kidney disease, the kidney develops abnormally, filling with large cysts that render it functionless. Scientists believe that multicystic kidneys develop due to a blockage in the ureter during early formation. When the kidneys begin producing urine, the urine has nowhere to drain and backs up, stretching the kidney into a functionless structure filled with grape-like cysts. Unlike polycystic kidney disease, multicystic kidney disease typically affects only one kidney. When the baby is born, the healthy kidney can take over for its cystic pair. But in rare instances when the cysts have invaded both kidneys, the baby is unlikely to survive much past birth.

Doctors usually discover multicystic kidneys in newborns via an ultrasound, then follow up with a CT scan to determine whether the cystic kidney is at all functional (in most cases, it is not). The image may reveal a kidney that has grown to eight to ten times normal size. If that is the case, and the enlarged kidney is putting pressure on other organs, the doctor may need to remove it surgically.

Polycystic Kidney Disease (PKD)

Polycystic kidney disease (PKD) is the most common life-threatening genetic disorder in the United States, affecting more than 600,000 people, according to the Polycystic Kidney Foundation (2001). In patients with PKD, a defective gene triggers the growth of multiple cysts in the kidneys. In one form of the disease, called autosomal recessive polycystic kidney disease (ARPKD), the cysts grow while the baby is still developing in the womb; in the other, called autosomal dominant polycystic kidney disease (ADPKD), the cysts grow over time, first appearing after about age 30 (which is why the disease is sometimes referred to as the "adult" form of PKD).

AUTOSOMAL DOMINANT POLYCYSTIC KIDNEY DISEASE (ADPKD)

Autosomal dominant polycystic kidney disease (ADPKD) is the most common form of polycystic kidney disease, accounting for about 90 percent of all cases. 1 in 400–1,000 adults have ADPKD, according to the Polycystic Kidney Foundation (2001). The disease is genetically dominant, meaning that if one parent has it, their child has a 50 percent chance of also having the disease. Unlike other genetic diseases, parents cannot carry ADPKD, but must suffer from the disease to pass it to their children. In rare cases, a genetic mutation can spontaneously trigger ADPKD in a person whose parents do not have the disease. Scientists do not know why this occurs, but they do know that once a person has ADPKD, he or she can then pass the faulty gene to subsequent generations.

Three ADPKD genes have been identified thus far, and each produces a different form of the disease. Scientists have located two of the three genes by isolating them on chromosomes, or strands of DNA. The first gene (called ADPKD1) is located on chromosome 16, the second (ADPKD2) is located on chromosome 4, and the third gene has yet to be found.

The genes responsible for ADPKD trigger production of an abnormal protein, which causes cells to reproduce at a greater-than-normal rate. These cells form the cysts. The cysts start out as small outpouchings or blisters in the nephrons of the kidneys. Most cysts detach from the nephron while they are still small (about one-eighth inch, or 2 mm, in diameter). As the cysts grow, they fill with either a clear liquid or a fluid containing red or white blood cells. The kidneys are forced to expand to accommodate the enlarged cysts. Whereas a normal kidney is about the size of a fist, a cyst-filled kidney can grow to the size of a football, and weigh more than 30 pounds.

A person who has ADPKD may live for years with the disease without realizing he or she has it, because the kidneys may work properly despite the existence of cysts. Often, the first symptom is high blood pressure, as cysts press on blood vessels within the affected kidney. High blood pressure affects between 60 and 70 percent of patients with ADPKD, according to the Polycystic Kidney Foundation (2001).

Following high blood pressure, the next symptoms to appear are usually blood in the urine and pain—which is located either in the back, in the side between the ribs and hips, or in the abdomen. The pain may be minor, or it may be so severe that it requires treatment with a procedure called cyst decompression, in which a doctor drains the fluid out of cysts with a needle, then coats the cysts with alcohol so they stop producing liquid and growing.

High blood pressure and pain are warning signs of ADPKD, but these symptoms may also be associated with a number of other diseases. Doctors don't usually suspect ADPKD unless the patient has a family history of the disease. The diagnosis is confirmed with an ultrasound. Cysts come in all different sizes—ranging from a tiny pinhead to a large grapefruit. Large cysts are easily visible on an ultrasound, but very small cysts may be difficult to spot. To view smaller cysts, doctors use either computed tomography (CT) or a more accurate, but very expensive, method in which family members who have ADPKD donate blood samples and the samples are compared with those of the patient.

People with ADPKD are more likely than the general public to suffer from urinary tract infections and kidney stones, because cysts can block the tubules of the kidneys and prevent urine from draining. Stagnating urine is a virtual breeding ground for infections and the chemicals that form kidney stones. ADPKD also creates a favorable environment for stone formation because it decreases the amount of citrate, a substance that inhibits stone formation, produced by the body. In addition to kidney stones and infections, patients with ADPKD are also more prone to intracranial aneurysms (a potentially fatal outpouching of blood vessels in the brain), groin and umbilical hernias, and diverticula (protrusions in the inner lining of the large intestine).

About half of ADPKD patients eventually develop kidney failure, but the likelihood of long-term kidney damage appears to be a factor of the genetic origin of the disease. People with the ADPKD2 gene tend to have a milder form of the disease and experience a slower progression to end-stage renal disease than those who have the ADPKD1 gene.

Although there is no cure for ADPKD, a number of treatments can slow cyst growth, prevent infection, and preserve kidney function.

High Blood Pressure Medication. If high blood pressure is left untreated, it can eventually damage the kidneys, enlarge the heart, and lead to strokes. Drugs such as angiotensin converting enzyme inhibitors (ACE inhibitors) bring down blood pressure, slowing the progression of kidney damage. Lifestyle changes such as diet, exercise, and reduced sodium intake can also help control blood pressure.

Antibiotics. When patients with ADPKD have a concurrent urinary tract infection, antibiotics prevent the infection from spreading to the cysts, where it becomes much more difficult to treat.

Pain Medication. While not a treatment, anti-inflammatory drugs such as acetaminophen may be necessary when the patient's pain becomes intolerable. Doctors don't recommend ibuprofen and aspirin because they can aggravate high blood pressure.

Surgery. Surgery is not a cure for ADPKD, but it can provide temporary pain relief. Laparoscopic surgery, a minimally invasive approach via a small incision, can effectively shrink large cysts.

Diet. Dietary changes can help damaged kidneys filter wastes more efficiently. Doctors recommend that patients eat fewer protein-rich foods (such as chicken, fish, and red meat), which force the kidneys to work harder. Drinking plenty of fluids helps the kidneys filter out wastes and prevents kidney stones. Avoiding salt can help control blood pressure. Preliminary research has indicated that caffeine may cause kidney cysts to grow faster, but no clinical studies exist to prove the benefit of avoiding caffeine.

Transplantation or Dialysis. If the kidneys fail, patients must resort to dialysis or a kidney transplant (see Chapter 5).

AUTOSOMAL RECESSIVE POLYCYSTIC KIDNEY DISEASE (ARPKD)

As in the dominant form of PKD, a genetic flaw is responsible for autosomal recessive polycystic kidney disease (ARPKD), but the gene is carried from generation to generation in a different way. Unlike ADPKD, parents may not actually have the disease themselves to pass it along, but may simply be carriers. If both parents carry the disease, their child has a 1-in-4 chance of developing ARPKD. The likelihood of having two parents with the disease is so slim that ARPKD only affects 1 in 10,000–40,000 people, according to the Polycystic Kidney Foundation (2001).

Symptoms of ARPKD can develop in the womb, in which case the malformed kidneys are unable to remove urine from the body, and the baby most likely dies at birth or soon after. If a baby survives infancy, he or she runs about a 50 percent chance of progressing to end-stage renal disease within ten years, which is why early detection is key. Some babies are actually diagnosed via ultrasound in utero. In other cases, children will visit their pediatrician with a urinary tract infection, frequent urination, and/or low blood cell count. Once the condition has been diagnosed, patients can preserve kidney function by taking medications that lower blood pressure (such as ACE inhibitors). Treatment also may include antibiotics to treat urinary tract infections, a nutritious diet to aid growth (because kidney function is crucial to normal growth, many children with kidney problems develop more slowly than their peers), and, finally, dialysis or transplantation if the kidneys fail.

FUTURE DIRECTIONS IN POLYCYSTIC KIDNEY DISEASE RESEARCH

Whereas once there was no hope for a cure, now researchers are learning what biochemical processes trigger PKD, and they are coming close to developing the first drugs to treat the disease. Scientists have located two of the three faulty ADPKD genes, and in early 2002 researchers at the Mayo Clinic in Rochester, Minnesota, pinpointed the gene responsible for ARPKD. Understanding which genes cause PKD will allow doctors to perform genetic testing and diagnose the disease earlier. Learning how these genes affect the chemical processes in the body that trigger cyst formation may eventually lead to a treatment or even a cure.

One of the fledgling treatments targets proteins in the body called growth factor receptors. Scientists believe that PKD overstimulates growth factor receptors, which leads to an overproduction of the cells that create cysts. Two drugs currently under investigation interfere with receptor activation to halt the production of cyst-forming cells. The first of these drugs, called EKI-785, was initially developed as a cancer treatment, but now researchers are studying its potential as a cyst inhibitor. In laboratory experiments, EKI-785 reduced the number of cysts, improved kidney function, and decreased mortality in mice. Researchers at Case Western Reserve University in Cleveland, Ohio, are experimenting with a similar growth factor receptor inhibitor, called EKB, which may be even more effective than EKI-785 in preventing cyst growth. Human trials will ultimately determine whether both drugs are safe and effective.

GLOMERULAR DISEASES

The kidney's filters, or glomeruli, may be tiny, but they carry the huge responsibility of filtering wastes out of the blood and into the urine. Any damage to the glomeruli from infection or disease can ultimately result in kidney failure. Two types of injuries hinder the glomeruli's ability to filter wastes: **glomerulonephritis** (see photo in color insert), an inflammation of the capillaries inside the glomeruli, and **glomerulosclerosis**, a scarring or hardening of the glomeruli. These are not diseases but conditions that result from the actions of diseases, infections, or, in some cases, medications.

Diseases That Cause Glomerulonephritis

INFECTION-RELATED GLOMERULAR DISEASE

When an infection strikes the body, the immune system immediately launches an attack, sending out proteins called antibodies to fight the invading virus. But in rare cases, that immune response backfires, and antibodies meant to destroy infection turn against the body. In the case of glomerular disease, the antibodies attack the glomeruli in the kidneys. Infection-related glomerular disease most often occurs in young children following an upper respiratory infection; viruses such as the chicken pox, measles, and malaria; or streptococcal infections of the throat or skin. Infections are responsible for the acute form of glomerulonephritis, which sets in quickly and eventually goes away on its own, as opposed to chronic glomerulonephritis, which may remain with a patient for life. Over 95 percent of patients with infection-related glomerular disease recover spontaneously and regain full kidney function within a few weeks. But in the few cases where kidney damage is permanent, dialysis or transplantation may be necessary.

LUPUS NEPHRITIS

Similar to infection-related glomerular disease, lupus nephritis is caused by an autoimmune response. In patients with this disease, the body deposits antibodies in the glomeruli that attack the tissue and cause inflammation. Twenty percent of patients with lupus nephritis progress to end-stage renal disease within ten years of onset, but treatment to suppress the immune response can prevent permanent kidney damage. Doctors commonly prescribe nonsteroidal anti-inflammatory drugs to reduce swelling, drugs that suppress the immune system (immunosuppressants) like prednisone, and cytotoxic drugs such as cyclophosphamide that control the overreacting lymphocytes behind the immune response.

GOODPASTURE SYNDROME

Doctors don't know what triggers this extremely rare autoimmune disease, but they believe it may follow a viral infection such as influenza. In goodpasture syndrome, an antibody attacks the kidneys and lungs. Lung damage is usually not long term, but kidney damage may be permanent, requiring dialysis or a transplant. Doctors treat goodpasture syndrome with a combination of immunosuppressive drugs, corticosteroids to control bleeding in the lungs, and a blood-cleaning therapy called plasmapheresis that filters out the harmful antibodies.

IGA NEPHROPATHY

The most common cause of glomerulonephritis, IgA nephropathy occurs when an immune response in the body deposits a protein called immunoglobin A (IgA) in the glomeruli, where it causes an inflammation. Doctors don't know exactly what causes IgA to land in the kidneys, but genetic factors may play a role, and often IgA nephropathy follows a recent bout of a respiratory or gastrointestinal infection. IgA prevents the glomeruli from filtering wastes, causing blood and protein to leak into the urine. The damage usually occurs slowly over time, so the disease may remain undetected for years. No real treatment exists for IgA nephropathy, but reducing dietary protein and controlling high blood pressure with medication can slow the progression of the disease. Studies have also shown that a daily dose of fish oil containing n-3 fatty acids, either alone or in combination with the immunosuppressant prednisone, may reduce inflammation and preserve kidney function.

Diseases That Cause Glomerulosclerosis

FOCAL SEGMENTAL GLOMERULOSCLEROSIS (FSGS)

Doctors don't know what causes FSGS, a disease that leads to progressive scarring of the glomeruli. The scarring damages the glomeruli to the point where they begin to leak protein into the urine. Once protein begins leaking, the disease goes into a vicious spiral. The leaking protein acts as a toxin, further damaging the glomeruli, until the kidneys are no longer functioning. No cure exists for FSGS, but treatment with immunosuppressants to control the immune response and ACE inhibitors to lower blood pressure can put patients in what is called "clinical remission."

MINIMAL CHANGE DISEASE (MCD)

MCD is a condition in which the glomeruli begin to leak large amounts of protein into the urine. This disease is most common in children, but it can also affect adults. Doctors don't know what causes MCD, but it may be associated with a history of immune disorders or with a recent vaccination or bee sting that triggered an immune response. The goal of treatment is to reduce the amount of protein lost in the urine, which can be done with a combination of corticosteroids and immunosuppressant drugs. Nonsteroidal anti-inflammatory drugs may also be effective in reducing proteinuria when steroids don't work.

DIABETIC NEPHROPATHY

Diabetes mellitus, described earlier in this chapter, elevates glucose levels in the blood, which increases the rate of blood flow through the kidneys. Over time, the pressure of so much blood flowing through the kidneys damages the glomeruli (see the section on diabetic kidney disease earlier in this chapter).

Nephrotic Syndrome

The result of long-term inflammation and scarring of the glomeruli is called nephrotic syndrome, a collection of symptoms in which the kidneys are damaged and unable to filter correctly. The injured kidneys leak large amounts of protein into the urine, resulting in a kind of malnourishment, in which the patient suffers from brittle hair and nails, weakened bones, stunted growth, and frequent infections. Many of the diseases associated with glomerulonephritis and glomerulosclerosis contribute to nephrotic syndrome, as well as a few others: cancers such as leukemia or lymphoma; infections like malaria, HIV, and hepatitis B; allergic reactions to insect stings, poison ivy, or snake venom; and reactions to certain drugs. As with many of the diseases in this category, there is no cure for nephrotic syndrome, but medications can keep the condition under control. Doctors currently use immunosuppressants, ACE inhibitors, and diuretics to reduce swelling.

Because protein leakage is often the most detrimental effect of glomerular disease, research efforts are focusing on a tiny cell called the podocyte, found in the Bowman's capsule of the glomeruli. Scientists believe that this cell is integral to the filtration process and is most likely at the root of protein leakage in diseases like MCD and FSGS. Researchers are trying to understand how the podocyte functions in order to derive new treatments based on these mechanisms.

OTHER KIDNEY DISEASES

Renal Tubular Acidosis (RTA)

The kidneys normally maintain a healthy acid-base balance in the body by excreting excess acids (hydrogen ions) into the urine. Normally the blood is slightly alkaline, but when the kidneys fail to remove enough acids, it becomes overly acidic (called acidosis). Because the body is always trying to maintain a balance between positively charged and negatively charged molecules (see Chapter 1), the excretion of one type of molecule can force other molecules out of the body as well. For example, too much acid in the body forces calcium from the bones and out of the system. Potassium, sodium, chloride, glucose, and phosphate may also be lost as the body struggles to maintain balance. Excess calcium and phosphate that wind up in the urine can form crystals, which can eventually develop into kidney stones.

Three distinct types of RTA exist, each varied in its causes and symptoms:

Type 1: Classic Distal RTA. "Distal" refers to the distal convoluted tubule, which is where the defect occurs. The distal convoluted tubule (see Chapter 1) is the furthest stretch of tubule in the glomerular capsule, through which substances not absorbed by the blood during filtration are sent to the ureter for excretion. In classic distal RTA, hydrogen ions are abnormally excreted from the distal tubule, causing the body to become overly acidic. Classic distal RTA can either be inherited or occur as a result of lupus, hyperparathyroidism, chronic urinary tract infections, or one of several other diseases. Classic RTA is also associated with increased potassium excretion. A lack of potassium, which helps regulate nerves and muscles as well as heart rate, can lead to weakness, cardiac arrhythmias, and paralysis. Calcium loss may stunt growth in children and cause progressive bone disease in adults. To control blood acidity and prevent leakage of calcium and potassium in patients with classic RTA, doctors prescribe sodium bicarbonate (baking soda) and sodium citrate.

Type 2: Proximal RTA. This less common form of RTA results from a defect in the secretion of hydrogen ions in the proximal convoluted tubule, where substances such as water, sugar, sodium, calcium, and proteins are normally reabsorbed into the blood (see Chapter 1). Proximal RTA occurs most often in children who have Fanconi's syndrome (a group of diseases marked by excess glucose, amino acids, citrate, phosphate, and potassium in the urine), but the condition can also result from inherited disorders that affect the way the body breaks down nutrients, for example cystinosis (a rare disease in which cystine crystals are deposited in the bones and tissues). Treatment for proximal RTA focuses on making the urine more alkaline with sodium bicarbonate or potassium citrate. Doctors may recommend that children with the condition take additional vitamin D to strengthen their growing bones.

Type 3. Type 3 RTA is a classification no longer used because it is now believed to be a combination of types 1 and 2.

Type 4: Hyperkalemic RTA. Like classic distal RTA, this condition is caused by a defect in the distal tubule, but in this case the problem results in elevated, rather than reduced, levels of potassium in the blood. Hyperkalemic RTA results from a deficiency of aldosterone, the hormone that regulates sodium, potassium, and chloride levels in the blood. In some cases, aldosterone is produced but the kidneys are unable to respond to the hormone. Several drugs have been known to cause hyperkalemic RTA, including spironolactone (used to treat congestive heart failure), ACE inhibitors, trimethoprim and pentamidine (antibiotics), heparin (used to stop blood from clotting), and nonsteroidal anti-inflammatory drugs. Doctors often prescribe sodium bicarbonate to correct acidosis, as well as medication to lower potassium levels in the blood.

Wegener's Granulomatosis

This extremely rare disease only affects 1 in 30,000–50,000 people, most of them adults, according to the Wegener's Granulomatosis Association (2001). Named for the German pathologist Friedrich Wegener (1907–1990) who first described the syndrome, Wegener's granulomatosis is characterized by an inflammation in blood vessel walls (vasculitis). The inflammation restricts blood flow and oxygen to various organs, including the kidneys. The disease usually starts out in the nasal passages and lung tissue, then attacks the respiratory system and kidneys. The first signs of the disease are cold-like symptoms (runny nose) that don't respond to medication. Wegener's may also affect the eyes (vision problems), ears (hearing loss), musculoskeletal system (joint pain), nervous system (paralysis), and skin (lesions). Nearly 80 percent of patients with the disease wind up with kidney damage.

Doctors diagnose Wegener's granulomatosis with blood tests to assess kidney function and detect the presence of ANCA, an antibody found in most Wegener's patients. The only definitive diagnosis is through a tissue biopsy, which is examined under a microscope for evidence of vasculitis and granulomas (a specific type of inflammation related to this disease). Without treatment, Wegener's can lead to kidney failure and other serious complications within a matter of months. But treatment with immunosuppressants and corticosteroids can usually put the disease into remission.

Renal Osteodystrophy

Healthy kidneys regulate the levels of calcium and phosphorous in the blood. Calcium strengthens bones, and phosphorous helps regulate calcium levels in the bones. But when the kidneys fail, too much phosphorous stays in the bloodstream, more calcium remains in the bones, and blood calcium levels drop. To restore balance, the parathyroid glands release parathyroid hormone, which draws calcium out of the bones and into the bloodstream. The result is one of several bone-weakening diseases known collectively as renal osteodystrophy. These include osteitis fibrosa, osteoclerosis, soft tissue calcifications, osteomalacia, and rickets (in children). Renal osteodystrophy is very common in patients with kidney disease and affects an estimated 90 percent of patients on dialysis.

Young children are especially at risk from renal osteodystrophy because their bones are still growing. Postmenopausal women are also vulnerable, because a drop in estrogen levels has made them prone to the degenerative bone disease osteoporosis. Often, bone damage occurs before the onset of any symptoms, which is why renal osteodystrophy is often called "the silent crippler." When symptoms (bone pain, joint pain, and muscle weakness) finally do appear, they may be minor or mimic those of other ailments. If left untreated, renal osteodystrophy can leave bones weak, brittle, and subject to easy fracture. In children, weakened bones can result in a deformity called renal rickets, in which the legs bend in toward the body or out away from the body. To prevent degeneration, doctors recommend that patients increase their daily calcium intake with supplements or injections of the hormone calcitrol, and eat fewer high-phosphorous foods such as milk, cheese, dried beans, and cocoa. Another effective treatment uses medications such as calcium carbonate (Tums) or calcium acetate (PhosLo) to decrease the absorption of phosphorous into the blood.

Renal Artery Stenosis

When the artery supplying blood to the kidney is narrowed or blocked, cells in the kidney are starved of the oxygen and nutrients they need to survive. The kidney responds to the lack of blood by releasing the hormone renin, which raises blood pressure. Narrowed arteries may be caused by arteriosclerosis (fatty deposits in the vessel walls—also known as hardening of the arteries), fibromuscular disease (fibrous tissue in the wall of the renal artery), or scar tissue following kidney surgery or injury. High blood pressure may be the only symptom of renal artery stenosis, and for this reason the disease may go undiagnosed until hypertension-related problems necessitate a visit to the doctor. Surgery or a balloon angioplasty (inserting a balloon-tipped catheter through the artery) are the methods most often used to clear the blockage.

DRUGS AND THE KIDNEYS (ANALGESIC NEPHROPATHY)

Just as disease can ravage the sensitive cells and filtering units of the kidneys, painkillers can also cause significant damage. Over-the-counter analgesics such as acetaminophen, aspirin, ibuprofen, and naproxen sodium are not normally injurious to the kidneys, but when taken by people predisposed to kidney problems, or when taken regularly for several years, these painkillers can do enough damage to result in end-stage renal disease. In rare cases, patients have suffered from acute kidney failure after fewer than ten days on painkillers. The risk is highest among patients who have systemic lupus erythematosus, a chronic kidney condition, or who have recently consumed large quantities of alcohol.

5

Kidney Failure: Dialysis and Transplantation

Kidney failure can be a gradual process, in which disease slowly wears down the tiny filters until they are no longer able to remove wastes from the body and regulate fluid balance. Or, the kidneys can suddenly grind to a halt following an infection or injury. Regardless of the cause, any disruption in kidney function can lead to the buildup of deadly toxins in the body, a rise in blood pressure, and a dangerous imbalance of fluid and chemical levels.

Acute kidney failure sets in rapidly following illness, injury, or any physiological change that acts upon the kidneys. For example, a patient might suffer serious muscle injuries during a car accident. When muscles are crushed, they release a molecule called myoglobin that blocks the kidney's tubules. Unable to filter wastes, the kidneys virtually shut down. Acute kidney failure may also follow major surgery if the patient suddenly loses a great deal of blood and goes into shock. That drastic drop in blood pressure prevents blood from reaching the kidneys for filtration. Any number of problems may trigger kidney failure: a bacterial infection, a blockage in the vessels that supply blood to the kidneys, excessive amounts of alcohol or other toxins in the system, dehydration from heat stroke or strenuous exercise, or an obstruction in the flow of urine.

The effects of acute kidney failure are rapid and often devastating. Waste and fluid removal grinds to a halt. Urine output may slow from the normal 1 to 2.5 quarts daily to less than a pint per day. Waste products, sodium, and potassium build up in the bloodstream. Trapped fluids cause the face, hands, and feet to swell. Those fluids also force the heart to work harder to pump a larger volume of blood through the body. In severe cases, acute kidney failure can lead to seizures, coma—and even death.

Doctors treat acute kidney failure by first addressing the underlying cause of the problem, for example by using antibiotics to clear up an infection or by surgically clearing a blockage. The kidneys can usually recover from short-term damage on their own, but patients are advised to limit fluids, and some may need dialysis to remove toxins during the healing period. Recovery may take anywhere from a few weeks to more than a year. Some people never fully recover from acute kidney damage and must deal with recurrent kidney problems for the rest of their lives.

In contrast to the rapid cessation of kidney function during acute kidney failure, chronic kidney failure develops gradually—so gradually, in fact, that many patients are not aware that they suffer from kidney disease until they have lost most of their renal function. Worn down by disease or damaged by overuse of certain medications, the kidneys eventually lose their ability to filter wastes and fluid from the blood. Chronic kidney failure is irreversible and inevitably leads to end-stage renal disease, a life-threatening condition in which the kidneys operate at only 5 to 10 percent of normal capacity.

The most common causes of chronic kidney failure are diabetes and high blood pressure, but a number of other diseases, drugs, and conditions (see Chapter 4) may also cause the kidneys to fail:

Kidney Diseases. Polycystic kidney disease, pyelonephritis (kidney infection), and glomerulonephritis (inflammation of glomeruli) all damage the kidneys' filters.

Analgesics. When taken over long periods of time, aspirin or nonsteroidal anti-inflammatory drugs such as ibuprofen and acetaminophen can inhibit kidney function.

Renal Artery Stenosis. Fatty deposits on the walls of the artery leading to the kidney (called arteriosclerosis or hardening of the arteries) block blood from reaching the kidney for filtration.

Toxins. Long-term exposure to fuels and solvents such as carbon tetrachloride or to the lead in paint, pipes, and soldering materials can lead to permanent kidney damage.

Obstructions. Stones and other blockages inhibit urine flow through the kidneys.

Chronic kidney failure often creeps in silently. The kidneys degenerate for months or even years with no symptoms. Even though the glomeruli are not able to filter as efficiently as they once could, they can still rid the body of enough wastes and fluids to compensate for lost function. But over time, as the kidneys lose more and more function and as fluids and wastes start to build up in the body, the warning signs become impossible to miss:

- Rising blood pressure
- Exhaustion, lightheadedness, or sluggishness from lack of oxygen in the blood as the kidneys produce less erythropoietin (EPO), the hormone necessary for red blood cell production
- Nausea from excess urea building up in the gastrointestinal tract
- Brittle bones, because the kidneys normally play a role in calcium absorption
- Achy joints from high levels of phosphorous in the blood
- Swelling in the eyes, arms, hands, and feet from excess fluids in the system

Kidney degeneration can occur quickly—within a few months—or the patient may live in relatively good health for several decades. Doctors try to sustain their patients for as long as possible by controlling the underlying cause of kidney failure (for example, reducing high blood pressure with medication) and by easing the kidneys' workload by restricting dietary intake of protein and fluids.

But, ultimately, the kidneys cannot hold out. Worn down by disease and overuse, their function will eventually diminish to only 5 to 10 percent of normal capacity. This is, as mentioned earlier in this chapter, called end-stage renal disease (ESRD). As of 2003, more than 375,000 Americans were being treated for ESRD; over half of them as a result of diabetes and/or high blood pressure, according to the National Kidney Foundation (2003). Glomerulonephritis and polycystic kidney disease also account for a significant proportion of ESRD cases. During the final stages of kidney failure, ESRD patients suffer from a number of serious health problems, including anemia, high blood pressure, bone disease, digestive tract problems, loss of mental functioning, and cardiovascular disease.

By the time a patient has reached the ESRD stage, medications and diet are of little use because the kidneys are unable to function on their own. Without dialysis or a kidney transplant, the prognosis is not good.

DIALYSIS

Once kidney function has slowed to below 10 percent of normal capacity, patients must rely on dialysis—an artificial filtering system that removes wastes, salt, and extra fluids from the body and regulates chemical balance. Dialysis can keep patients reasonably healthy until a kidney becomes available for transplant, but it is not a cure for kidney failure. Survival rates during the first year of dialysis surpass 75 percent, but after five years, only about 30 percent of patients are likely to survive without a transplant.

There are two methods of dialysis: **hemodialysis**, which involves the use of a special filtering machine, and **peritoneal dialysis**, which uses the patient's abdominal cavity as a filter.

Hemodialysis

Hemodialysis has been available since the 1960s, and it is currently the most common dialysis method used in the United States. When a patient undergoes hemodialysis (see Figure 5.1), his or her blood is pumped through tubes into an artificial kidney, called a dialyzer. The machine has two parts: One side holds the blood, and the other contains dialysate fluid. A thin, semipermeable membrane separates the two parts. Blood passes through microscopic holes in the membrane and into the dialysate solution. The solution contains special chemicals that act like a sponge, pulling wastes and excess fluids from the blood. Meanwhile, electrolytes and other essential chemicals pass from the dialysate solution into the blood. Protein molecules and red blood cells, which are too large to fit through the membrane, remain in the blood. The cleansed blood flows back through another set of tubes into the patient's body. Patients usually visit a hospital or dial-

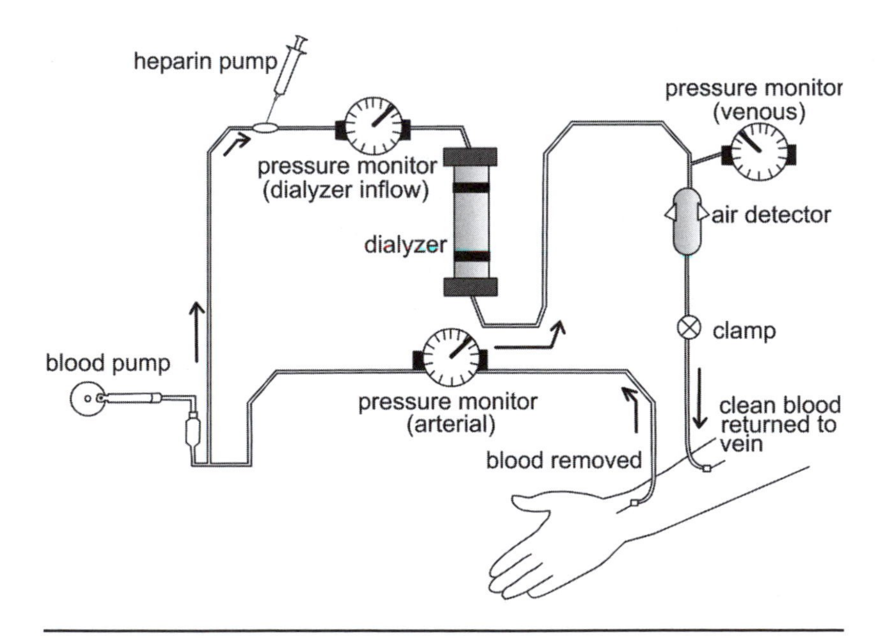

Figure 5.1. The hemodialysis process.
The patient's blood is pumped through an artificial kidney, called a dialyzer. The filtered blood is returned to the patient via a vein.

ysis center three times a week and spend about four hours hooked up to a dialyzer (though newer, more efficient high-flux hemodialysis machines can shorten the session time). In some parts of the country, hemodialysis can be done at home with the help of a family member or friend who has been trained in the technique.

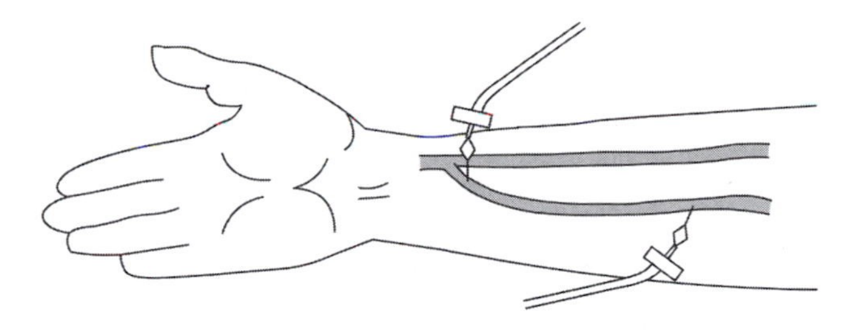

Figure 5.2. Fistula.
An artery in the patient's forearm is connected to a vein. The increased blood flow strengthens the vein so that it can withstand repeated needles during the hemodialysis process.

Several months before treatment begins, doctors must create an access through which the dialysis machine can be connected to the patient's body, via a fistula or graft. The preferred method is a fistula (see Figure 5.2), in which a surgeon connects an artery, usually in the patient's forearm, directly to a vein. Increased blood flow from the artery strengthens the vein so that it can withstand repeated needle stickings. If the patient's blood vessels won't accommodate a fistula, doctors connect the artery to the vein via a synthetic tube, called a graft (see Figure 5.3). The disadvantage to a graft is that it remains outside of the body, where it is more prone to infection than a fistula. Grafts are also more susceptible to blood clots and may need to be replaced sooner than fistulas.

If a patient's kidney disease has progressed so quickly that he or she requires immediate dialysis, doctors may insert a catheter (see Figure 5.4) directly into a vein in his or her neck, chest, or leg to provide temporary fast access. The catheter has two chambers to allow the blood to flow in and out of the dialysis machine. Catheters are usually used as a short-term solution,

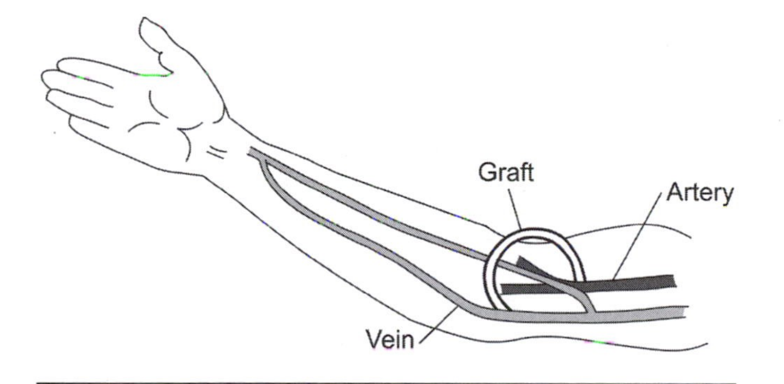

Figure 5.3. Graft.
Similar to the fistula, but the patient's artery is connected to the vein via a synthetic tube.

Figure 5.4. Catheter.
When immediate dialysis access is required, a catheter is inserted into the patient's neck, chest, or leg. The two tubes allow blood to flow in and out of the dialysis machine.

but they may be implanted under the skin if needed for longer periods of time.

About once a month, a patient is tested to make sure that the dialysis is removing enough waste from the bloodstream. The two most common tests monitor blood urea nitrogen (BUN) levels to identify whether impurities remain in the system. In the first test, called urea reduction ration (URR), doctors compare the levels of urea in the blood before and after a dialysis session. If the dialysis is working, the URR should measure at least 60 percent. The second test, called Kt/V (Kt stands for the dialyzer clearance multiplied by time, and V represents the volume of water in the patient's body), is derived from the URR, but it also takes into account the urea generated by the body and the extra urea removed during a dialysis session. This test is a more accurate measure of how much urea the dialysis session actually removed. If Kt/V values measure below 1.2, patients will need to be on dialysis for longer periods of time or use a more efficient dialyzer that filters the blood more quickly.

Peritoneal Dialysis

Peritoneal dialysis (see Figure 5.5) has been in use since the 1980s, but recent advancements have made the technique more effective and have eliminated some of the side effects. The advantage to this procedure is that it offers greater mobility, because the dialysis is done right in the patient's own body. That means patients can conduct dialysis sessions at home, or on vacation for that matter, without having to visit a special center or hospital.

Before the first dialysis session, doctors surgically implant a soft tube called a catheter into the patient's abdomen. During each session, the patient uses the catheter to fill the abdomen with dialysate solution. The solution contains the sugar dextrose, which pulls wastes and excess fluid from the blood into the abdominal cavity through blood vessels in the peritoneal membrane (the membrane surrounding the digestive organs). Dialysate remains in the abdomen for a four- to six-hour period called the dwell time.

Once all of the wastes and excess fluids have been removed from the blood, the dialysate is drained from the abdomen and replaced with a fresh solution. The process of draining and filling the abdomen, called an exchange, takes between thirty and forty minutes. The dwell times and number of exchanges needed on a daily basis differ from patient to patient, depending on how fast an individual's peritoneal membrane can filter. On average, a peritoneal dialysis patient will undergo about four exchanges a day.

The three types of peritoneal dialysis are continuous ambulatory peritoneal dialysis (CAPD), continuous cycling peritoneal dialysis (CCPD), and intermittent peritoneal dialysis (IPD).

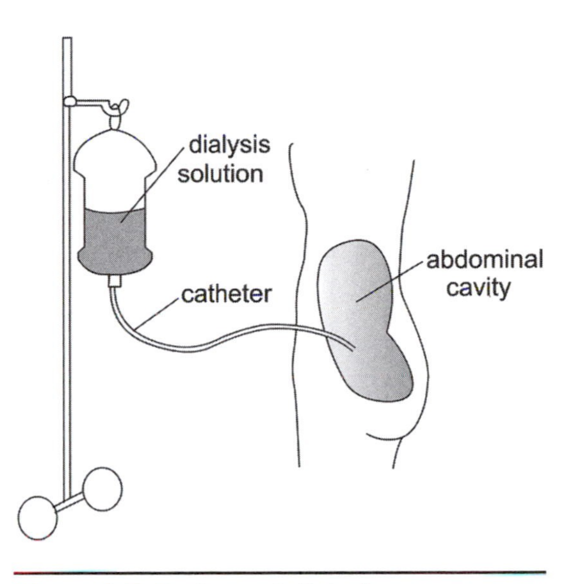

Figure 5.5. Peritoneal dialysis.
The patient's abdomen is filled with dialysate solution via a catheter. Filtration occurs in the peritoneal membrane. After a 4- to 6-hour dwell time, the solution containing filtered wastes and fluids is removed from the patient's body.

CONTINUOUS AMBULATORY PERITONEAL DIALYSIS (CAPD)

The most common form of peritoneal dialysis, CAPD uses gravity, rather than a machine, to fill and empty the patient's abdomen with dialysate solution. Most people who use CAPD change their dialysis solution at least four times a day and sleep with the solution inside their abdomen. The advantage to this technique is that it can be performed at home and does not require the patient to be hooked up to a machine, allowing greater mobility and comfort.

CONTINUOUS CYCLING PERITONEAL DIALYSIS (CCPD)

This method uses a cycler machine to automatically infuse and remove the dialysate solution. The majority of CCPD exchanges are done at night rather than during the day. While the patient sleeps, the machine cycles fluid in and out of the abdominal cavity for between three and five cycles, each lasting about 1.5 hours. In the morning, the patient begins one cycle that lasts the entire day. Some patients may require an additional exchange in the afternoon hours when the morning cycle is not adequate to remove all of the wastes and fluids. Some people, especially those who weigh more than 175 pounds, require a combination of CAPD and CCPD for more efficient waste removal.

INTERMITTENT PERITONEAL DIALYSIS (IPD)

IPD is performed only three to four times a week, but for longer sessions—between ten to fourteen hours at a time. In a variation of this process, called nocturnal intermittent peritoneal dialysis (NIPD), the exchanges take place at night rather than during the day. This method is usually reserved for patients with some remaining kidney function or whose peritoneal cavities are especially efficient at filtering wastes.

Doctors periodically evaluate their patients to make sure that the peritoneal dialysis is working effectively. The two tests most commonly used to measure waste removal are:

> *Peritoneal Equilibration Test (PET).* Measures how much sugar has been removed from a dialysis solution and how much urea and creatinine have entered the solution during a four-hour exchange.
>
> *Clearance Test.* Compares the amount of urea in the used solution to the amount remaining in the patient's blood.

If the dialysis process is not adequately removing wastes, the doctor may increase the number of exchanges, increase the volume of solution infused into the peritoneum with each exchange, add another night exchange for CAPD patients, or add another day exchange for CCPD patients.

Complications Associated with Dialysis

Dialysis is considered safe, but like any medical procedure it has potential side effects. One of the most common complications is an infection at the hemodialysis access site. Any time the body is left open to the air, it becomes susceptible to a bacterial attack. Infection in peritoneal dialysis patients, called peritonitis, occurs at the site where the catheter is connected to the body. Doctors usually treat any type of dialysis-related infection with antibiotics. Hemodialysis patients are also susceptible to blood clots, because doctors must interfere with the arteries and veins to create an access site.

Patients who are on dialysis for five years or more are prone to a condition called dialysis-related amyloidosis. As mentioned in Chapter 4, patients who have amyloidosis produce and store an abnormal amount of protein fibers, which cluster together and build up in the kidneys. In dialysis-related amyloidosis, the body is unable to properly remove a protein called beta-2-microglobulin. The molecules of this protein band together, forming large deposits in the kidneys, bones, joints, and tendons.

Managing Diet in Dialysis Patients

By the time patients require dialysis, their kidneys are only working at about 10 percent of capacity. Even though dialysis removes wastes and excess fluids from the body, it cannot fully take over hormone production and electrolyte balance from the kidneys. For this reason, dialysis patients have to be especially careful about what they eat and drink, and may require vitamin supplements and special medications to stay healthy.

Limit Protein. Higher-protein diets increase the amount of urea produced by the body and force the kidneys to work harder to remove it. Some protein is necessary for maintaining muscle strength, but doctors usually recommend that dialysis patients limit the amount of meat, poultry, fish, and eggs in their diet to keep waste production low.

Cut Down on Foods High in Potassium and Phosphorous. Potassium, which is found in bananas, oranges, vegetables, chocolate, and nuts, plays a role in regulating the heartbeat. The kidneys normally remove excess potassium from the body, but without their control, potassium levels can reach dangerous heights. Too much phosphorous in the body is also a problem, because in large amounts, phosphorous draws calcium from the bones, leaving them fragile and brittle. Doctors recommend that dialysis patients hold back on phosphorous-rich foods such as milk, cheese, nuts, dried beans, and dark colas. Patients may also need medication to control phosphorous levels in their blood, especially between dialysis sessions.

Drink Less Fluids. Normally, the kidneys would filter out extra fluid into the urine. But because dialysis patients must rely on an artificial removal process, which may not be done every day, they need to watch what they drink. Too much liquid in the system causes the body to become bloated, raises the blood pressure, and forces the heart to work overtime.

Avoid Salt. Sodium not only causes the body to retain water (see Chapter 1), but it also increases thirst.

Vitamin Supplements. Having to comply with all of the dietary restrictions mentioned here means missing out on a number of important vitamins. Dialysis patients sometimes need a vitamin supplement to replace missing dietary nutrients. Because their diet is so limited, hemodialysis patients may need to increase their caloric intake by adding vegetable oils, honey, jam, or sugar (of course, sugary foods are not recommended for patients with diabetes). Conversely, peritoneal dialysis patients may actually have to count their calories during treatment, because the dialysis fluid contains extra calories.

EPO and Iron Supplements. As mentioned earlier in the chapter, many patients with kidney failure suffer from anemia, or a low volume of red blood cells. This problem worsens as kidney failure progresses, so that by the time patients require dialysis, they will most likely need treatment for anemia as well. Anemia results from a lack of the hormone erythropoietin (EPO), normally produced by the kidneys. It can also be triggered by blood loss during dialysis, as well as by low levels of iron and folic acid. Dialysis patients may receive a genetically engineered form of EPO intravenously during treatment, along with iron injections and possibly folic acid supplements.

Future Directions in Dialysis Research

Scientists are conducting a number of studies that will one day help them improve dialysis treatments and reduce the likelihood of complications. One National Institutes of Health (NIH) trial is attempting to tackle problems associated with grafts and fistulas for hemodialysis access. About a quarter of dialysis-related problems that require hospital stays stem from the vascular access site, many of them due to thrombosis, or blood clots.

Researchers are currently evaluating antithrombotic agents to reduce or even eliminate the risk of blood clots.

Other trials are investigating better and more efficient dialysis filters, reusable hemodialysis membranes, faster high-flux machines, and new methods for returning essential substances to the blood during dialysis while removing impurities.

KIDNEY TRANSPLANT

To a patient who has been on dialysis for years, the words "We've found you a kidney" are welcomed with the greatest relief imaginable. A kidney transplant can mean freedom from dialysis tubes and needles, freedom from restrictive diets, and freedom to travel without being tied to a machine. More important, a transplant can mean the difference between life and death. Survival rates decline rapidly after five years on dialysis, but kidney transplant success rates over the same period of time top 90 percent. The trouble is that human kidneys are not easy to come by, and dialysis patients often spend years on a waiting list, hoping for news that a transplant will soon be available. Doctors perform about 13,000 kidney transplants every year in the United States, but there are currently more than 50,000 people on the waiting list for a kidney, and that number is rising every year, according to the United Network for Organ Sharing (2003). Sadly, many patients die before ever receiving a donor kidney.

Where Do Kidneys Come From?

Most of the kidneys transplanted in the United States come from cadavers, many of whom have been declared brain dead after an accident. Kidneys may also come from living donors—usually a friend or family member of the patient. Living donors are always preferred, because their kidneys result in higher survival rates. The average lifespan of a living donor kidney is fifteen to twenty years, as compared to seven to ten years for one from a cadaver. When it comes to live donors, blood relatives are always preferred over spouses or friends, because they are more likely to genetically match the recipient. The chance of finding a "perfect match" is about 25 percent from a sibling, as compared with 6.5 percent from a nonliving donor.

Patients who need a kidney but don't know anyone willing to donate are placed on a United Network for Organ Sharing (UNOS) transplant waiting list. Once on the list, patients are assigned points based on how well they match available kidneys, how long they have been on the list, the urgency of their medical condition, and their age. The average wait time has been rising over the past decade. In 1999, more than 60 percent of patients had been on the list for more than one year, and the average wait time for a kidney was about three years.

When a kidney becomes available, information about the donor is sent to the UNOS's central computer, which generates a list of potential recipients. Surgeons remove the donor's kidney, while the potential recipients go through a full regimen of medical tests to see if they are potentially a good match. The typical medical evaluation includes:

- A complete medical history and physical exam, to assess the patient's overall health.
- Blood tests to determine whether the patient's blood type matches that of the donor. Blood type is the most crucial match in the evaluation process. If the patient and recipient do not have the same blood type, the transplant will not be successful. The patient's blood is also tested for infectious diseases, such as hepatitis, HIV, syphilis, Epstein-Barr virus, and cytomegalovirus (CMV), any of which would preclude a transplant.
- Histocompatibility testing to see if the donor and recipient's tissue types are compatible. Human cells carry six markers, called human leukocyte antigens (HLAs), that distinguish tissue type. Each of us inherits three markers from our mother and three from our father. The more markers that match between donor and recipient, the less likely that the new kidney will be rejected. A perfect match is called a six-antigen match and is most likely to occur among family members. But as long as the donor and recipient's blood types match, a six-antigen match is not required for the transplant to proceed.
- Cross-matching, in which a small sample of the donor's blood is mixed with a sample of the recipient's blood in the laboratory. The presence of antibodies in the recipient's blood could be a sign that his or her body will reject the transplanted kidney. No reaction is a good sign that the transplant will be a success. Another test, called the mixed leukocyte culture, mixes white blood cells, or leukocytes, from the donor and recipient. Leukocytes protect the body against invading bacteria, viruses, and parasites. If the donor and recipient cells react to one another, the two are not a likely match.
- Panel reactive antibody (PRA) test measures immune system activity. When the immune system is on high alert, the potential for rejection increases. When the immune system is at its calmest, the transplant is more likely to be successful.
- Electrocardiogram (EKG or ECG) to determine how well the patient's heart is working and to look for signs of an undiagnosed heart disease.
- Pulmonary function test to assess lung strength and determine how well the blood is able to carry oxygen.
- Upper gastrointestinal (GI) and lower GI examinations to look for signs of disease or other abnormalities in the stomach and intestines.
- Ultrasound to check the strength of the iliac blood vessels, which will be attached to the new kidney's blood vessels.
- Cancer tests, which may include a pap smear and mammogram for women and a prostate exam for men.

Once doctors have established a match, the kidney is shipped to wherever it is needed and the surgery (see photo in color insert) can begin. The new kidney is placed beneath the patient's malfunctioning organs in the lower abdomen and attached to the patient's blood supply through the iliac artery and a vein. To allow urine flow out of the kidney, surgeons attach it to the bladder through the ureter via a small plastic catheter. The old kidneys are rarely removed (unless they are infected or filled with cysts), because that would involve major surgery. Over time, the old kidneys atrophy, or get smaller, from lack of use. After a period of days or weeks, the new kidney slowly begins to take over for the diseased organs, producing urine and regulating the body's chemical and fluid balance.

Patients usually remain in the hospital for about a week after surgery, then visit with their doctor regularly for follow-ups to make sure the transplanted organ is working properly. During some of these visits, doctors may do an ultrasound of the new kidney to check that blood vessels are providing it with an adequate blood supply. Doctors may also remove a small sample of tissue from the kidney, called a biopsy, to check for signs of rejection or other complications. Finally, they may do a CT scan or MRI to view the kidney from different angles and look for infections, fluid collection, or other problems.

The key to a successful transplant, regardless of how closely donor and recipient are matched, is suppression of the patient's immune system to avoid rejection of the new kidney. The patient will need to take antirejection drugs, for example cyclosporine A, for as long as he or she has the kidney. Advancements in the development of immunosuppression drugs have meant vastly improved survival rates (see Table 5.1 for survival rates) for transplant patients. But still, drugs may not be enough to prevent a patient's body from rejecting what it sees as a foreign invader. Rejection may occur suddenly (acute) or over time (chronic). Doctors look for a dramatic rise in the patient's creatinine level, especially in the first year after transplant, which would indicate that the new kidney is under attack. If the kidney is rejected, the patient must return to dialysis until doctors can locate a new organ. Kidney recipients are also susceptible to infections such as cytomeg-

TABLE 5.1. Kidney Transplant Survival Rates

Years	Cadaver Transplant	Living-donor Transplant
1	94.8%	97.6%
3	88.9%	94.6%
5	81.8%	91%

Source: UNOS, 2002.

alovirus, herpes simplex 1 and 2, fungal infections, and yeast infections, all of which are treated with antibiotics.

If the transplant is a success, patients can expect to resume their normal activities, free from the burden of dialysis treatments. Diet regulations are usually relaxed, but patients still need to watch their fluid and salt intake (which can increase blood pressure).

Future Directions in Transplantation Science

Kidney transplants can prolong survival and restore quality of life for most ESRD patients. The problem is finding enough viable kidneys to save every patient waiting for a replacement organ. The demand for organs has prompted significant research into other possible sources, among them xenotransplantation, or the use of animal organs. The idea has been met with fierce debate: Some doctors say it is unethical to outfit human beings with animal organs, especially considering the increased risk of rejection, but others counter that the practice could save thousands of lives each year. One animal considered as a potential donor is the pig, which is similar in anatomy and physiology to humans, and can be genetically modified during breeding to promote even greater similarity.

Aside from the ethical dilemmas involved in xenotransplantation, scientists must combat the violent and almost immediate rejection with which the organs would be met in human patients. Animal-to-human transplants may trigger what is called hyperacute rejection, in which the patient's blood coagulates in the new organ, killing it almost immediately. Animals also carry infectious diseases that could easily spread to humans, especially among patients whose immune system is being suppressed to prevent rejection. For these reasons, xenotransplantation is still years away from practical implementation. (See "The Future of Organ Transplantation.")

Another promising, albeit even more controversial, approach to organ procurement, involves the use of stem cells to generate entire kidneys, livers, or other organs. Stem cells, found in the tiny forming embryo, are the building blocks of every tissue, every muscle, and every organ in the human body. Using stem cells harvested from an embryo created out of a patient's cells and a donated egg, scientists are hoping to one day grow replacement organs in the laboratory. In 2002, researchers at Advanced Cell Technology in Worcester, Massachusetts, announced that they had grown miniature kidney-like organs using cells from cloned cow embryos. When the kidneys were implanted into the cow that had donated the cells, they were able to filter wastes and make urine. Scientists have made great strides toward therapeutic cloning, but the technology is still in its infancy, and the debate over harvesting organs and tissue from human embryos has raged from church pulpits to the U.S. Congress. Even scientists are divided over whether the technique's merits outweigh its ethical dilemmas.

The Future of Organ Transplantation

The idea of humans outfitted with animal parts sounds like the stuff of science fiction, but it may soon be a reality thanks to a new breakthrough in genetics. Scientists have long been searching for a way to fill the growing need for replacement organs. Thousands die every year while waiting for organ transplants, because there are never enough human donors to fill the need. Animal organs, however, are in plentiful supply. The problem is that they are difficult to transplant in humans because of the likelihood of severe rejection and infection.

Those problems may be overcome in the foreseeable future, thanks to a breakthrough technology. In the summer of 2002, PPL Therapeutics, the Scottish company that helped clone the sheep Dolly, announced the birth of four healthy piglets. What made the birth announcement so significant was that the pigs were born without the two genes that normally cause humans to reject pig organs. The scientists had removed, or "knocked out," these genes in a laboratory.

The next step is to test genetically altered animal organs on other animals, and then on people. Researchers are hopeful that "knock out" and other new technologies may make xenotransplantation a reality in the twenty-first century.

In the more foreseeable future, doctors are looking at new methods by which to tackle the problem of rejection following organ transplant surgery. One technique under investigation is called peripheral blood stem cell transplantation, a treatment currently used in cancer patients. Doctors inject immune cells from the kidney donor's blood into the recipient in the hope that the cells will register as friend rather than foe and prevent the recipient's immune system from attacking the foreign tissue. Another method is a blood-filtering process called plasmapheresis, coupled with a drug that suppresses immune system antibody production. The technique prevents the patient's body from producing antibodies that would otherwise attack the transplanted organ. The goal is to eventually eliminate the need for antirejection drugs, which may cause serious complications and leave the patient susceptible to infection. Doctors are also looking closely at a procedure that filters antibodies out of a kidney patient's blood before surgery, thus eliminating the need for an exact blood and tissue match.

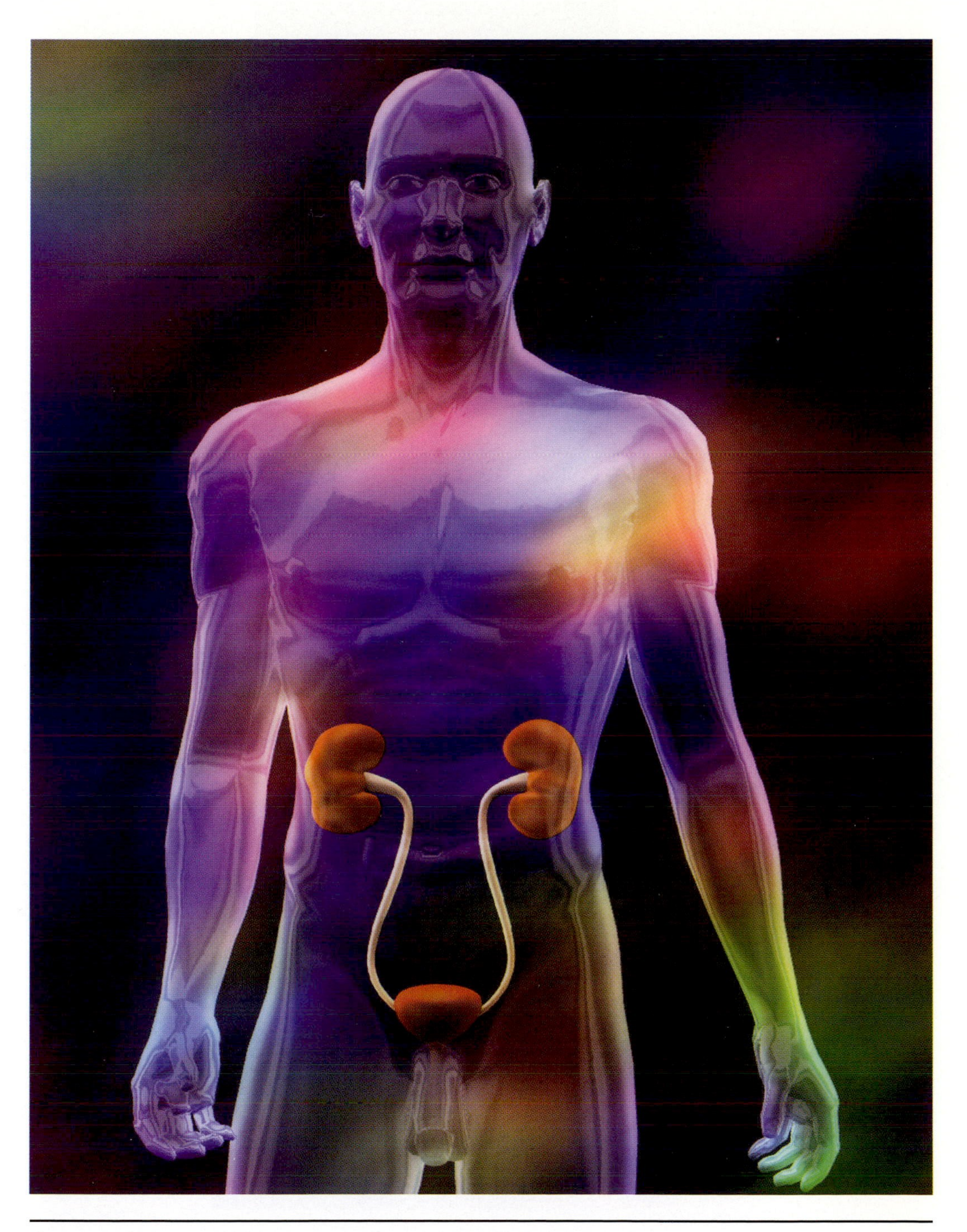

Computer illustration depicting the kidneys and bladder isolated in their proper positions in the human body. © H. Singh/Custom Medical Stock Photo.

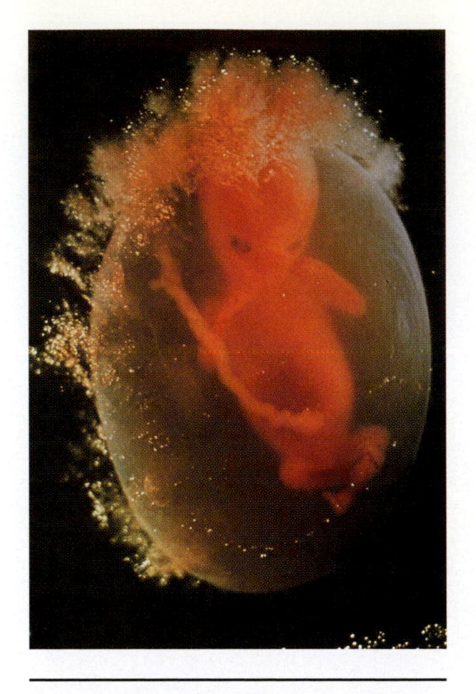

Human fetus at 9 weeks. © Collection CNRI/Phototake.

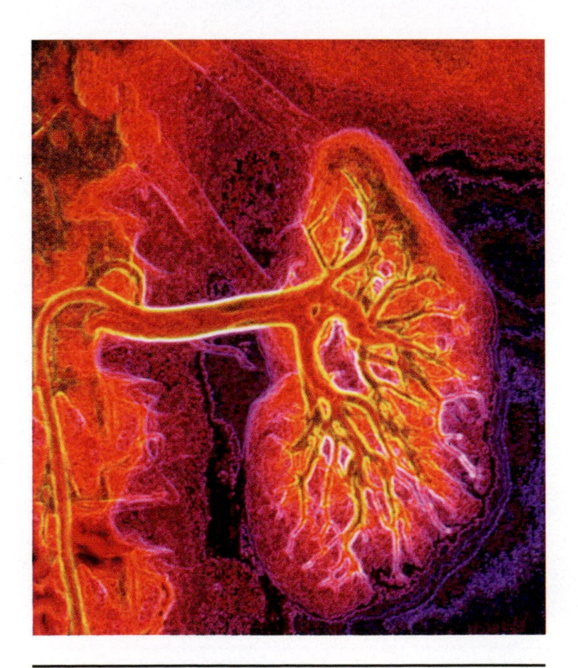

X-ray showing normal vascularization of the left kidney (frontal view). © ISM/Phototake.

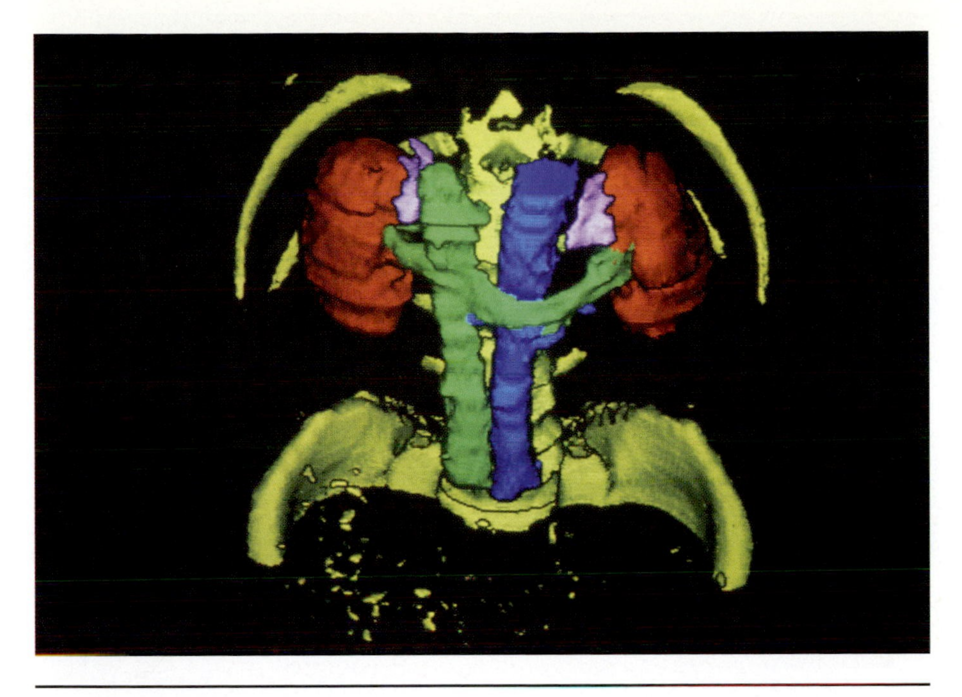

3D CAT scan of the kidneys, the renal artery, the renal vein, and the adrenal glands. © Collection CNRI/Phototake.

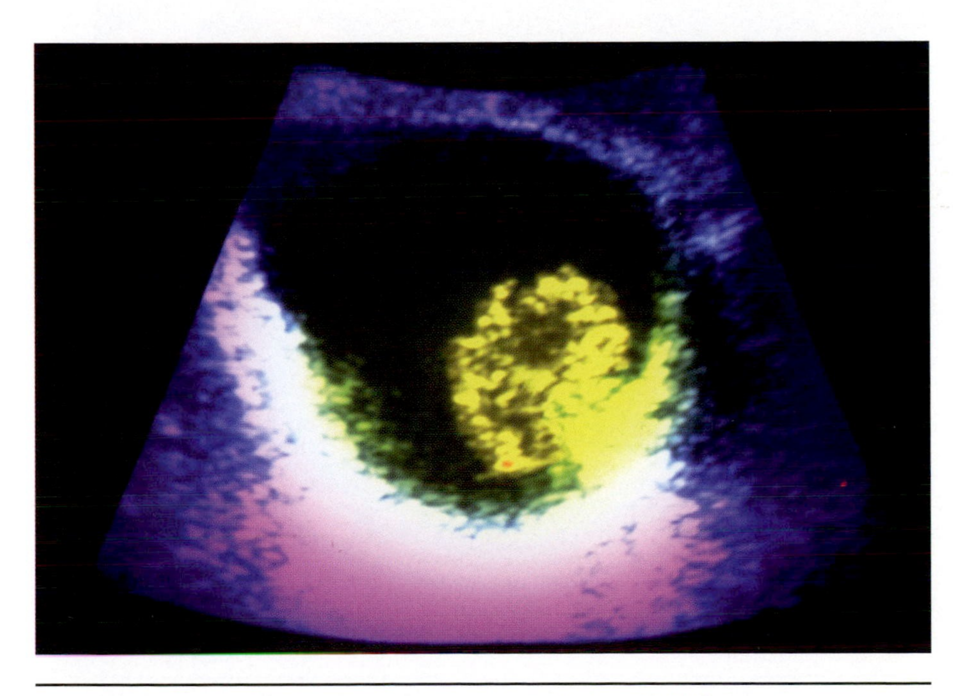

Ultrasound scan showing large bladder tumor. © Collection CNRI/Phototake.

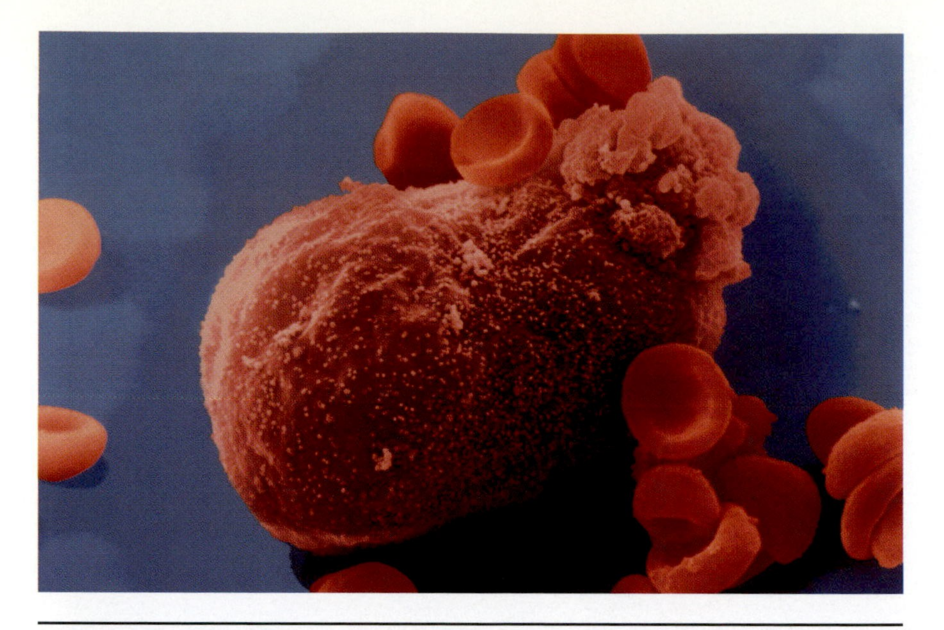

Colorized scanning electron micrograph of cancer cell from kidney. © Jean Claude Revy–ISM/Phototake.

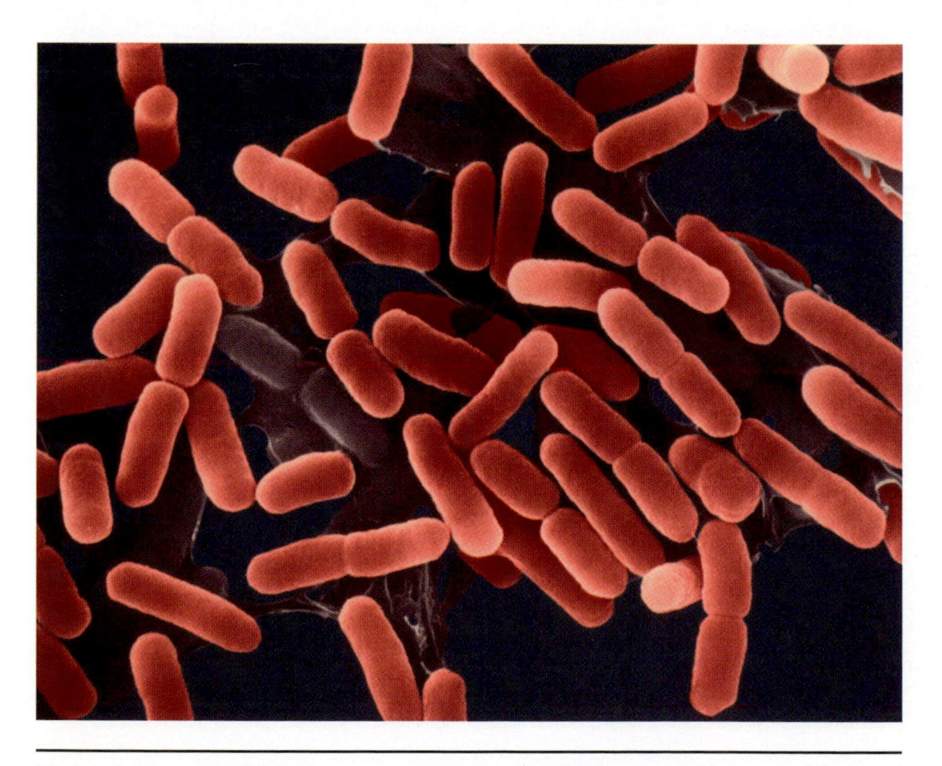

Most urinary tract infections are caused by the *E. coli* bacteria, which is normally found in the gut. © Dennis Kunkel/Phototake.

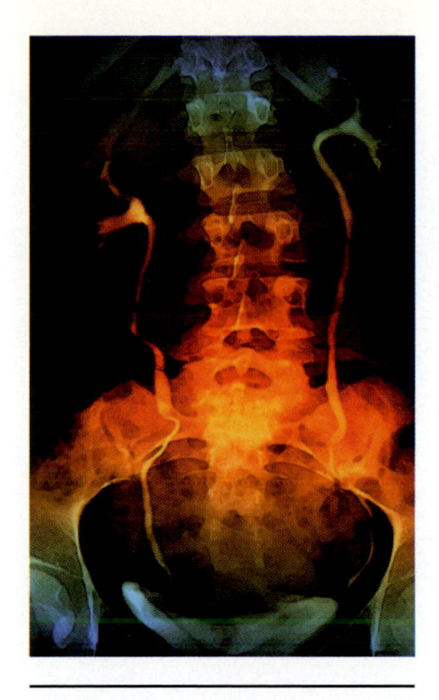

Intravenous urogram (x-ray of the urinary tract) front view, showing normal urinary tract (renal pelvis, ureters, bladder). © ISM/Phototake.

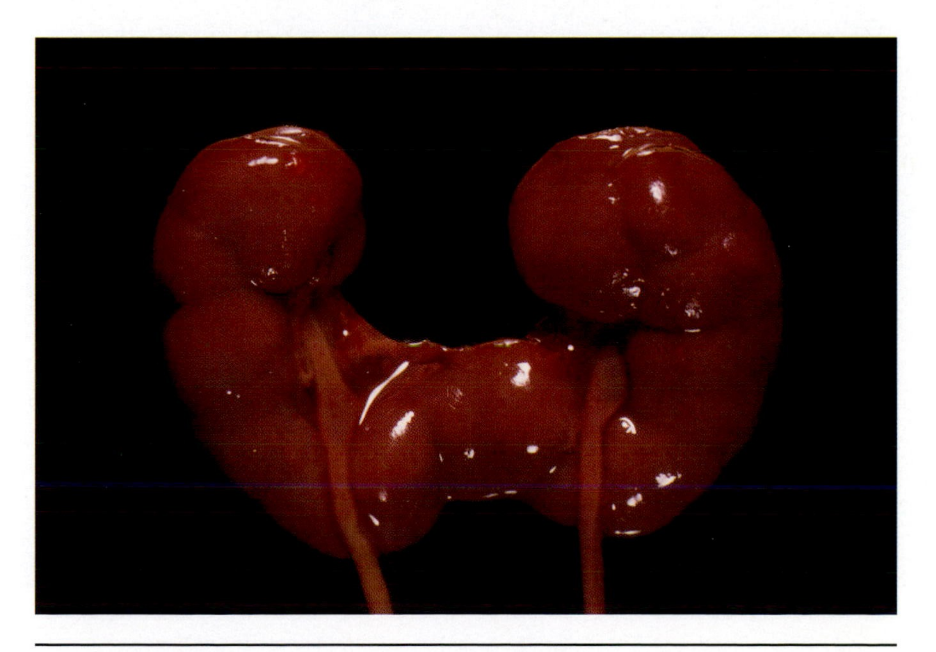

A horseshoe kidney. The two kidneys have fused together at the base, forming a "U" shape. © J. Siebert/Custom Medical Stock Photo.

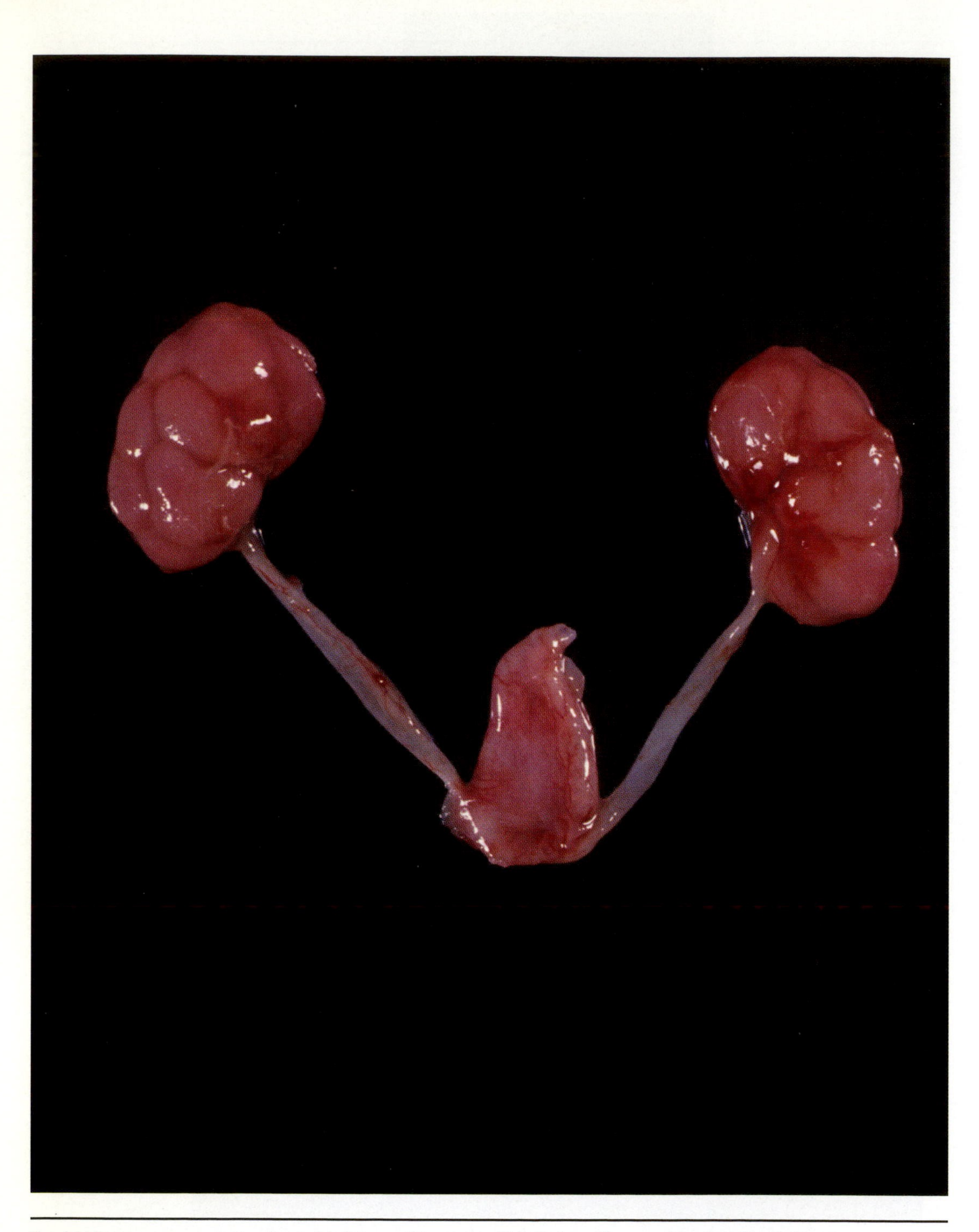

Fetal kidneys, ureters, and urinary bladder at 21 weeks. © J. Siebert/Custom Medical Stock Photo.

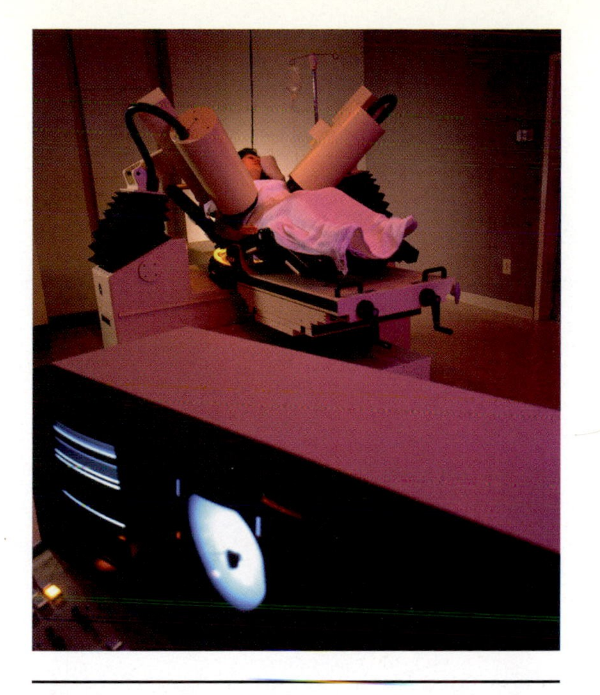

Woman undergoing lithotripsy, or kidney stone removal. Lithotripsy crushes the stone into small enough fragments to be removed from the body. © Yoav Levy/Phototake.

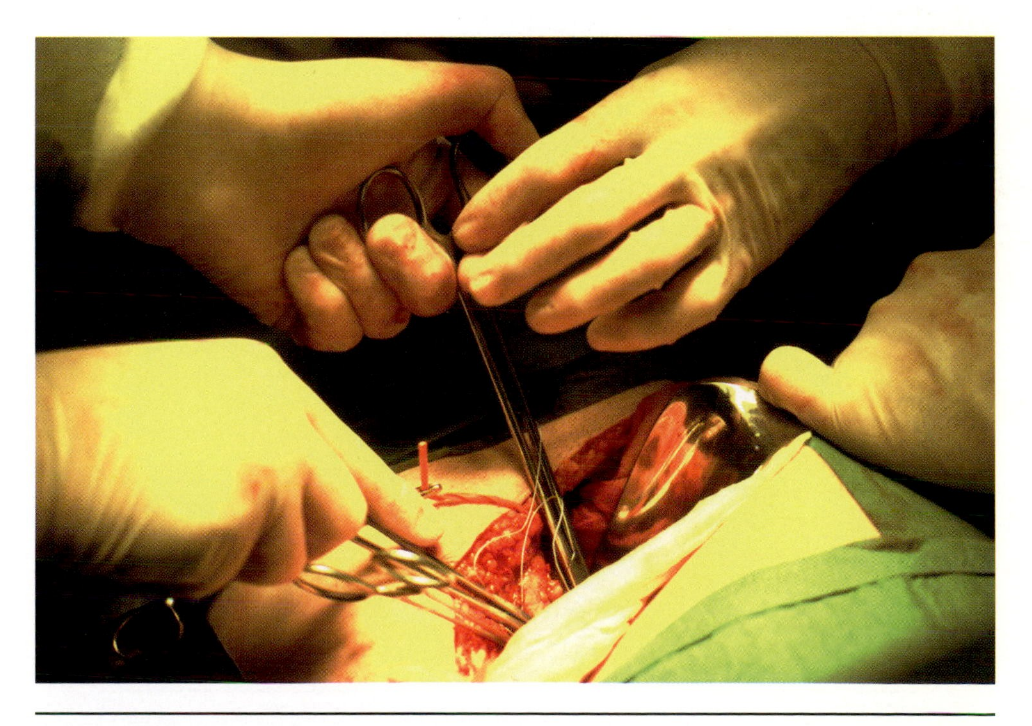

Kidney transplant operation, from mother to daughter. Kidneys donated by family members are preferred, because they are more likely to genetically match the recipient. © Index Stock/Phototake.

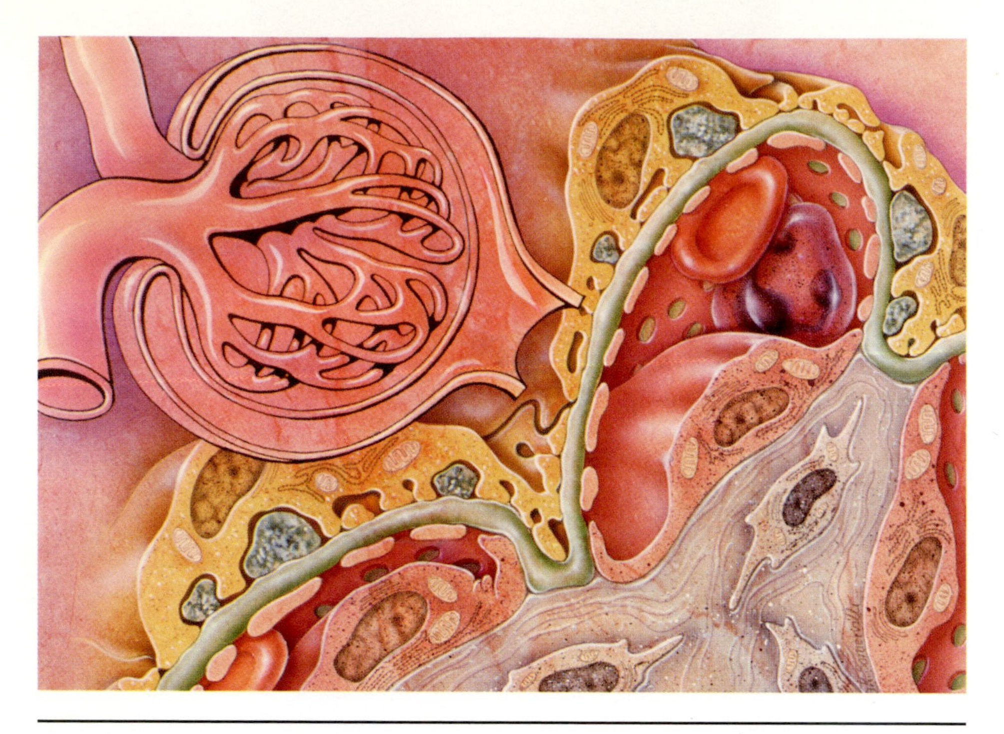

Glomerulonephritis, inflammation of the kidney. © Kevin A. Somerville/Phototake.

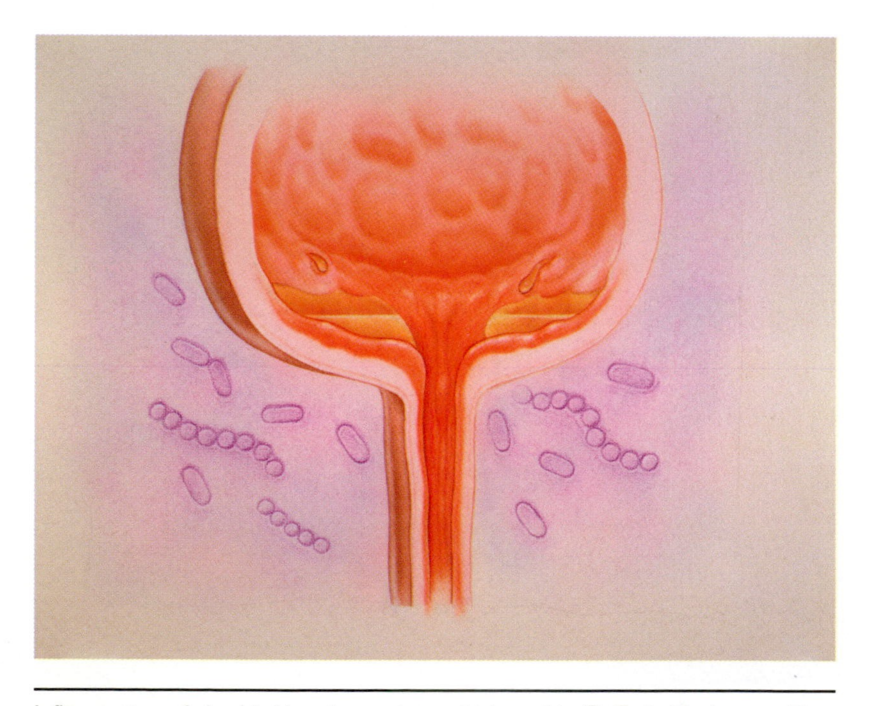

Inflammation of the bladder due to interstitial cystitis. © Craig Zuckerman/Phototake.

6

Diseases and Disorders of the Bladder

The bladder is designed to work as a receptacle—a place to store urine produced by the kidneys until it is removed from the body via urination. In a normal urination cycle, the detrusor muscle in the bladder wall relaxes, allowing the organ to stretch and accommodate an increasing volume of liquid. When the bladder is full, receptors in its walls send a message to the spinal cord via sensory neurons. The spinal cord passes the message along to the brain in the form of impulses. When a message is sent back that it is time to release the urine, the bladder muscles contract to push urine through the urethra and out of the body (for more on the process of urination, see Chapter 1).

A number of conditions can affect the bladder's ability to hold and properly release urine. The bladder may be overactive, releasing urine too often or at inopportune times; or it may be underactive, failing to contract until the urine finally overflows and pushes back up into the kidneys (more on this subject in Chapter 8). The bladder may become infected or inflamed, causing pain and a frequent or urgent need to urinate. Or, the bladder may come under attack by an army of cancerous cells.

BLADDER CANCER

Although bladder cancer does not garner the type of media attention generated by breast, lung, and other cancers, it nonetheless poses a serious health threat. According to the American Foundation for Urologic Disorders, the disease affects more than 50,000 Americans each year. Bladder cancer is the fourth most common cancer among men and the ninth most common among women in the United States. It most often strikes the elderly, with the highest rates of incidence among people in their 70s. For some unknown reason, race and gender are important factors in bladder cancer susceptibility. Caucasians are twice as likely to develop the disease as African Americans or Hispanics, and men are three times more likely than women to suffer from bladder cancer.

Lining the wall of the bladder is a layer of cells that protects the sensitive bladder tissues from toxins in the urine. The two types of cells that comprise this layer are called transitional cells and squamous cells. These cells normally replicate only when needed, but sometimes they begin to grow uncontrollably until they form a tumor. Cancerous, or malignant, tumors differ from benign tumors in that they can spread and have the potential to threaten healthy tissues throughout the body (for more on tumors, see the discussion of kidney cancer in Chapter 4). Almost all bladder cancers are transitional cell carcinomas, so called because they originate in the transitional cells of the bladder lining. Transitional cells also line the inside of the kidney, ureters, and urethra, which is why most bladder cancers look almost identical to transitional cell carcinomas elsewhere in the urinary tract and may actually recur in another part of the urinary system after they are removed from the bladder.

Only about 8 percent of bladder cancers are squamous cell carcinomas; that is, they form in the squamous cells of the bladder. An even smaller percentage of bladder cancers are adenocarcinomas. This type of cancer is usually seen in high-risk patients who were born with a rare congenital abnormality called bladder exstrophy (see the section on congenital bladder abnormalities later in this chapter), or who have had tuberculosis of the bladder. Cancer may also spread to the bladder from other organs, but it retains the name of the organ from which it spread (for example, if it originated in a woman's cervix, the cancer would still be named cervical cancer).

Risk Factors for Bladder Cancer

SMOKING

Smoking is the single greatest risk factor for bladder cancer. Smokers are twice as likely to develop bladder cancer as nonsmokers, according to the American Cancer Society (2003). When carcinogens contained in cigarette, pipe, or cigar smoke move from the lungs into the bloodstream, they continue on through the kidneys, where they are filtered into the urine. The cancer-causing chemicals damage the protective cells lining the bladder, increasing the risk of a malignancy. How much a patient smokes, what type of tobacco he or she uses, and how deeply he or she inhales all play a role in determining the risk of developing bladder cancer.

CHEMICAL EXPOSURE

People who are regularly exposed to a class of chemicals called arylamines run a significantly increased risk of developing bladder cancer. Individuals who work in the rubber, chemical, metal, leather, printing, painting, dye, or textile industries may come into contact with these carcinogenic chemicals on a daily basis. Studies have found that hairdressers who regularly work with permanent chemical dyes face a fivefold risk of bladder cancer, and that women who use these dyes have double the risk of women in the general population.

CHRONIC BLADDER INFLAMMATION

Recurrent urinary tract infections and kidney and bladder stones, all of which irritate the lining of the bladder, have been linked to an increased risk of bladder cancer (especially squamous cell carcinoma). But although patients with chronic infections and stones may be more prone to bladder cancer, these conditions don't necessarily cause the disease.

MEDICATIONS

Ironically, drugs used to treat certain types of cancer may sometimes increase a patient's risk of developing bladder cancer. Cyclophosphamide, used to slow malignant cell growth, is just one of several anticancer medications attributed to an increased bladder cancer risk.

FAMILY/PERSONAL HISTORY

An individual whose mother, sister, or other immediate family member has had bladder cancer runs a slightly higher risk of developing the disease than someone without a family history. Also, a majority of people who have survived bladder cancer will have to deal with the disease again at some point in their lives. Bladder cancer recurs in up to two-thirds of patients.

BLADDER EXSTROPHY

A rare congenital defect that occurs when the bladder fails to close properly in the developing fetus leaves the bladder exposed to chronic infection, which can eventually lead to the formation of bladder adenocarcinomas.

Symptoms of Bladder Cancer

Bladder cancer is often difficult to detect because it often presents with no symptoms in its early stages. The first sign may be blood in the urine (hematuria), which is sometimes too small to be seen with the naked eye (microscopic hematuria) (see Chapter 4 for a discussion on hematuria). Blood cells shed in larger amounts can turn the urine a pale yellow-red or a darker red. Other symptoms may include frequent urination or urgency

What Does a Urologist Do?

A urologist is a doctor who specializes in the diagnosis and treatment of urinary tract diseases. Included in a urologist's repertoire are disorders related to the kidneys, bladder, urethra, and prostate. He or she may also treat penile dysfunctions and reproductive issues (although many patients choose to visit an infertility specialist, rather than a urologist, for the treatment of infertility).

Urogynecologists are specialists within the field of urology who treat women for incontinence and disorders of the pelvic floor. Pediatric urology specialists treat children for everything from bedwetting to congenital malformations of the urogenital tract.

Urology evolved out of the ancient art of stonecutting. Practitioners hundreds of years ago traveled around Europe curing patients of bladder stones with scalpels and other rudimentary instruments. Today's urologists have much more sophisticated equipment at their disposal. They may use x-rays, CT scans, and a host of other scanning technologies to diagnose their patients, as well as lasers, lithotripsy, microwave energy, and surgical scopes to treat disorders.

(the need to urinate without the ability to do so). Because these symptoms are indicative of several urinary system diseases, including kidney stones, bladder infection, or a benign tumor, doctors must run a battery of tests to confirm a bladder cancer diagnosis, and may refer their patients to a urologist (a specialist in urinary system diseases) for a more thorough examination. (See "What Does a Urologist Do?")

Diagnosing Bladder Cancer

PHYSICAL EXAM AND MEDICAL HISTORY

The doctor will first run through the patient's personal and family history of bladder cancer, and ask about any risk factors such as smoking or occupational exposure to cancer-causing chemicals. The next step is a physical exam, during which the doctor will feel the patient's abdomen and pelvis for any lumps or hard masses.

URINE TESTS

A urine culture can rule out conditions that have the appearance of bladder cancer. The doctor will examine the patient's urine under a microscope to look for blood or bacteria (signs of a urinary tract infection). Another option is a cytology test—viewing the urine under a microscope to look for cancerous cells. The test may be conducted on a regular urine sample or by taking a **bladder washing**, in which a saline (salt) solution is placed inside the bladder and then removed for testing. This procedure is very effective in detecting high-grade (more advanced) tumors, but it may miss low-grade tumors. Cytology also has a high false-positive rate; that is, it may indicate cancer when there is none. Newer urine tests, which look for markers or signs of cancer in the urine, can reveal the recurrence of bladder cancers earlier and more effectively than more invasive techniques such as cytology and cystoscopy (more on biomarkers later in this chapter).

CYSTOSCOPY

Using a thin, lighted scope inserted through the patient's urethra into the bladder, the doctor can examine the bladder interior from different angles. The procedure is usually performed either in the doctor's office or in an outpatient facility. If the doctor finds a tumor, he or she can remove a tissue sample using instruments inserted through the cystoscope. That tissue would then be examined under a microscope by a pathologist, who can determine whether it is indeed cancer and, if so, how far it has spread (the stage, or progression, of the cancer).

INTRAVENOUS PYELOGRAM (IVP) OR INTRAVENOUS UROGRAPHY (IVU)

An IVP, otherwise known as an IVU, is similar to a traditional x-ray, with the exception that contrast dye is first injected into the patient's blood vessels. The dye outlines the kidneys, ureters, and bladder so that the doctor may clearly view them on x-ray film. The technique is effective at revealing larger tumors or ruling out kidney stones or other urinary tract disorders, but a cystoscopy is usually necessary before the doctor can confirm a bladder cancer diagnosis.

COMPUTED TOMOGRAPHY (CT) SCAN

Using a special machine that takes x-rays from various positions around the patient's body, a CT scan creates a cross-sectional image of the urinary tract to reveal whether the cancer has spread. The CT scan can pick out large tumors; however, it may miss smaller tumors, and it is unable to distinguish how far a tumor has advanced.

MAGNETIC RESONANCE IMAGING (MRI) SCAN

Similar to the CT scan, an MRI produces cross-sectional images of the urinary system to allow the doctor to see whether the cancer has spread from the bladder to nearby tissues, lymph nodes, or distant organs. But unlike the CT scan, an MRI uses powerful magnets and radio waves to create this image.

ULTRASOUND

During pregnancy, an ultrasound uses high-frequency sound waves to display an image of the growing fetus on a video screen. The technique has a similar application in bladder cancer detection (see photo in color insert). Ultrasound is primarily used to stage bladder cancers before surgery.

OTHER TESTS

If the doctor suspects that the cancer has spread to other organs, he or she may also call for bone scans, liver function tests, or chest x-rays to seek out the malignancies.

The procedures outlined here, along with a tissue biopsy, can confirm a diagnosis and help the doctor determine how far the cancer has progressed (stage) and how aggressive the cancer is likely to be (grade). Five stages are used to diagnose bladder cancer. The higher the stage and grade, the more dour the prognosis.

> *Stage 0, or carcinoma in situ.* The earliest stage of bladder cancer is considered superficial, because the cells lie only on the surface of the inner bladder lining and have not spread to the muscle or tissue of the bladder wall. This is the least invasive and easiest type of bladder cancer to treat.
>
> *Stage 1.* Cancer cells have penetrated the inner tissue lining the bladder, but have not reached the muscular wall.
>
> *Stage 2.* Cancer cells have spread to the inner lining of muscles, but have not passed completely through the muscles.
>
> *Stage 3.* Cancer cells have spread through the bladder wall to the layer of tissue surrounding the bladder. Once cancerous cells have spread outside the bladder, they are usually found in nearby lymph nodes. From the lymph nodes, the cancer can easily make its way to other organs. In stage 3, the cancer may have spread to, but no farther than, the nearby reproductive organs—the uterus or vagina in women, or the prostate in men.
>
> *Stage 4.* The most invasive form of bladder cancer. Cancerous cells have progressed through the wall of the abdomen and have most likely spread to lymph nodes and to faraway organs like the lungs.

Treatment Options

The doctor will specially tailor a patient's treatment to fit his or her type of bladder cancer. Superficial tumors carry a better prognosis and can usually be removed via simple surgery. More invasive tumors may require removal of the entire bladder, in combination with radiation or chemotherapy.

TRANSURETHRAL RESECTION (TUR)

Superficial cancers that have not progressed past the bladder lining may be removed via **transurethral resection (TUR)**, a relatively simple surgical procedure. Using a **resectoscope** (a specialized version of the instrument used in cystoscopy) fitted with a small wire loop at the end, the doctor removes the tumor from the bladder lining and burns away any remaining cancer cells with high-energy electricity (called **electrofulguration**). After removing the tumor, the doctor can look at a tissue sample under a microscope to determine how aggressive the cancer is and how deeply into the bladder lining it has spread. Side effects of the procedure are typically minor and temporary, and may include blood in the urine or pain during urination. Four out of five patients can be cured using TUR alone, says the American Foundation for Urologic Disease. But for the small percentage of people with more invasive bladder cancer, one or more of the following treatments may be necessary.

CYSTECTOMY

In this procedure, doctors remove all or part of the bladder. A partial cystectomy involves removal of just the affected area of the bladder. This procedure works well in patients with localized cancer. Patients with more invasive cancers will usually require complete bladder removal, along with nearby lymph nodes, part of the urethra, and often nearby organs. In men, these may include the prostate, seminal vesicles (glands that produce the fluid portion of semen), and part of the vas deferens. In women, the surgery may remove the uterus, ovaries, fallopian tubes, and part of the vagina.

Bladder removal raises a serious health concern. Where will the urine be stored? To solve this problem, doctors create a urinary diversion by one of several methods. The preferred technique, called an ileal conduit, attaches a piece of the patient's small intestine (ileum) to the ureters at one end and to an opening in the wall of the abdomen called a stoma (see Figure 6.1) at the other end. A bag fits over the opening to collect urine. Doctors may also create a reservoir internally, but the patient must then drain the urine through a catheter. A newer, and often preferable method of urinary diversion uses part of the small or large intestine to create a new bladder, which is then attached to the urethra. The urine passes out of the body through the urethra just as it would during natural urination.

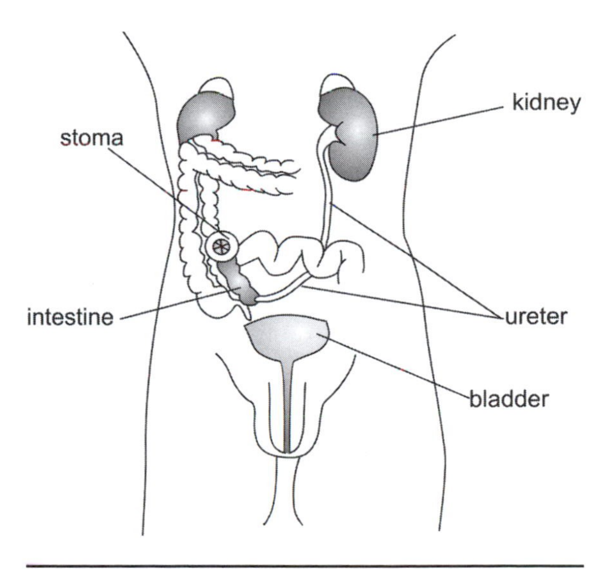

Figure 6.1. Urinary diversion after cystectomy (bladder removal) procedure. To divert the urine after the bladder has been removed, a portion of the patient's small intestine is connected to the ureter at one end, and to an opening in the wall of the abdomen called a stoma at the other end.

In addition to the problems of rerouting urination, patients who undergo a radical cystectomy must often deal with sexual side effects caused by the removal of reproductive organs. Women whose ovaries and uterus are removed will no longer be able to become pregnant. If part of the vagina is removed, intercourse may be painful or even impossible. At one time, nearly all men who underwent radical cystectomy became impotent, but surgical improvements now help most men retain sexual function.

CHEMOTHERAPY

These special drugs, which are designed to kill cancer cells anywhere in the body, are used to combat invasive cancers that have spread beyond the bladder. Chemotherapy may be used on its own or in conjunction with surgery or radiation therapy. Research has found that, when used before surgery, chemotherapy nearly doubles survival rates in patients whose cancer has spread to the bladder wall but has not yet invaded other organs. Chemotherapy can be administered in pill form, injected through a vein, or placed directly into the bladder through a process called intravesical chemotherapy.

Because chemotherapy can kill healthy cells as well as cancerous cells, it leaves patients susceptible to a number of side effects. Patients may be more prone to infections, may bruise or bleed easily, and often experience a lack of energy. Other side effects include nausea, hair loss, vomiting, and loss of appetite.

RADIATION THERAPY (RADIOTHERAPY)

This procedure targets high-energy x-rays to kill cancer cells and shrink tumors. Unlike chemotherapy, which kills cells throughout the body, radiation only affects specific areas. This technique is often used after surgery to kill any cancerous cells that remain. In patients with advanced bladder cancer, for whom surgery is not an option, radiation may ease pain and bleeding. Radiation can be directed at the patient's body from the inside or outside. External radiation therapy uses a machine to aim radiation at the tumor from the outside. Internal radiation places a radioactive substance inside the patient's bladder, which is removed after a couple of days. Some patients require a combination of external and internal radiation therapies. Patients usually receive five radiation treatments a week for five to seven weeks. Spreading out the treatments over twenty-five or more sessions protects healthy cells by reducing the amount of radiation exposure in one sitting.

Side effects depend on the size of the dose and duration of the treatment, but may include drowsiness (especially during the latter weeks of the treatment), a sunburn-like redness in the treated area, nausea, vomiting, urinary discomfort, and an increased risk of infection. These side effects are usually temporary.

IMMUNOTHERAPY (BIOLOGICAL THERAPY)

Immunotherapy uses the body's own immune system to fight cancer. Drugs called biological response modifiers, which are either made by the body or recreated in a laboratory, stimulate white blood cells in the immune system. One method for administering the biological response modifiers is by placing a solution into the bladder via a catheter. Doctors often use BCG (bacillus Calmette-Guerin) solution for bladder cancer immunotherapy. BCG contains a weakened form of bacteria used in the tuberculosis vaccine. The solution remains in the bladder for about two hours, and the procedure is repeated once a week for about six weeks. Immunotherapy is most often used to prevent a recurrence in superficial cancers that have been treated with transurethral resection (see the beginning of this section). Side effects may include bladder irritation, increased urinary urgency or frequency, pain during urination, low-grade fever, nausea, and/or blood in the urine.

Most patients are successfully treated with one or a combination of the therapies discussed here. But bladder cancer is notorious for making an unwanted reappearance in the bladder, kidneys, or ureters. Regardless of the treatment, about 80 percent of bladder cancers recur within five years, so doctors must be vigilant about following up on their patients. Every three months for at least a year, patients must return to their doctor for a cystoscopy, which checks for the emergence of new tumors. The follow-ups will also include routine urine cytologies (checking for cancer cells under the microscope) and yearly IVPs. Patients will require yearly follow-up examinations for the rest of their lives.

Future Directions in Bladder Cancer Research

BIOMARKERS FOR DIAGNOSIS

Cystoscopy remains the preferred method for detecting bladder cancer, but it is an invasive and expensive procedure. Cytology is less invasive, but it is also less sensitive in picking up low-grade tumors. Scientists are currently investigating markers in the urine, which they believe can detect a recurrence of bladder cancer months before the tumors actually appear. The procedure is noninvasive, inexpensive, and extremely sensitive in picking up even low-grade cancers. Two urine marker tests are currently approved by the U.S. Food and Drug Administration (FDA) for finding recurrent bladder cancers after treatment: Bladder tumor antigen (BTA) detects proteins in urine that are produced when bladder tumor cells reproduce, and nuclear matrix protein (NMP 22) measures the levels of certain proteins released from the nuclei of tumor cells. Patients with bladder cancer have about twenty-five times greater levels of NMP 22 than healthy people.

Other markers currently under investigation are telomerase, enzymes released by most cancer cells but rarely by healthy cells; and survivin, a protein that inhibits apoptosis (cell death). Cancer patients typically have an excess of survivin, because it keeps tumor cells alive and allows them to continue multiplying. Biomarkers are not expected to replace cystoscopy, but rather to reduce the frequency with which it is needed. Urine marker tests could, in the future, be used along with cytology to more effectively monitor patients for a bladder cancer recurrence.

PHOTODYNAMIC THERAPY

In this procedure, doctors inject a chemical into the patient's blood that becomes activated when exposed to light. Once the cancer cells absorb the drug, the doctor shines a special laser light inside the bladder through a cystoscope. The light induces chemical changes in the drug that cause it to attack cancer cells but spare healthy cells. Photodynamic therapy is only effective in treating the early stages of bladder cancer, because the light can only reach cells on the surface of the bladder. Side effects are usually mild, and may include skin irritation and increased sensitivity to sunlight.

GENE THERAPY

Like immunotherapy, gene therapy better equips the patient's body to fight off cancer cells. Doctors inject bladder cancer cells with a virus containing a special gene, which inhibits the growth of cancer cells.

VITAMIN THERAPY/CHEMOPREVENTION

People who have bladder cancer are at increased risk for developing cancer of the kidneys, ureters, urethra, or another part of the bladder. Vitamin therapy and chemoprevention bolster the body's immunity with large doses of drugs or vitamins to prevent a recurrence after treatment.

ANTIANGIOGENIC THERAPIES

This procedure "starves" tumors by blocking the blood vessels that supply them with nutrients and oxygen. Developed by Judah Folkman, a Harvard University cancer researcher, antiangiogenic therapies are currently under investigation as a potential cure for cancer. The chemicals used in this procedure, which may include endostatin, angiostatin, or thalidomide, are far less toxic than radiation and chemotherapy treatments.

INTERSTITIAL CYSTITIS

Although doctors have been aware of interstitial cystitis (IC) for the better part of a century, they remain mystified as to what causes this painful bladder disorder and are often helpless to treat it. Many patients are still accused of inventing their symptoms or are misdiagnosed with a urinary tract infection and prescribed an ineffective dose of antibiotics. About the only thing that is known for sure is that IC is much more common than was once believed. According to the Interstitial Cystitis Association, about 700,000 Americans live with the condition, the vast majority of whom are women.

In patients with IC, the bladder wall becomes inflamed and irritated (see photo in color insert). The inflammation can eventually scar the bladder, interfering with the organ's ability to stretch and accommodate normal volumes of urine. Another hallmark of the disorder is pinpoint bleeding on the walls of the bladder, called **glomerulations**. Unlike other urinary diseases, IC does not progressively worsen, but patients may go through periods during which their pain intensifies and their bladder loses some of its capacity.

Unlike simple cystitis, interstitial cystitis is not caused by bacteria. IC patients normally test negative for bacteria on urine cultures and do not respond to antibiotic therapy. Several theories exist as to where the disorder originates. Some doctors believe that IC is triggered by an autoimmune response following a bladder infection. That is, antibodies produced by the immune system to protect the body turn against it and attack the bladder. Another theory is that IC patients have a substance in their urine that irritates their bladder, but that substance has yet to be discovered. Heredity may also play a role, but research has uncovered no gene to confirm this suspicion. Finally, some doctors believe that IC is caused by a defect in the bladder lining that allows bacteria and other toxic substances in the urine to irritate the bladder wall.

Most researchers believe that IC is not one but two diseases. About 10 percent of patients have the ulcerative form, characterized by Hunner's ulcers (named for the doctor who discovered them) in the bladder. The other 90 percent of IC patients have the nonulcerative form of the disease.

Symptoms of Interstitial Cystitis

Symptoms can vary from patient to patient, but generally bladder inflammation causes pain or pressure in the pelvic or lower abdominal area, an increased need to urinate (urgency), a more frequent need to urinate (frequency), or a combination of urgency and frequency (more on these symptoms in Chapter 8). The pain may increase as the bladder fills and empties, or during sexual intercourse, because of the close proximity of the bladder to the sexual organs. Some patients complain of irritable bowel-like symptoms (abdominal cramping and bloating) accompanied by diarrhea or constipation. Because the inflamed bladder can no longer hold as much urine, people with severe cases of IC may urinate as frequently as sixty times a day.

Diagnosing Interstitial Cystitis

Because IC is similar in appearance to several other urologic, gynecologic, and intestinal illnesses (including UTIs, bladder cancer, bladder inflammation caused by radiation, endometriosis, STDs, kidney stones, bacteria in urine, or prostatitis in men), diagnosis is often a process of ruling out other disorders that have the same symptoms. Once other illnesses have been eliminated as possibilities, doctors look for three characteristic symptoms of IC:

Glomerulations. As the bladder stretches during filling, tiny blood vessels that resemble the kidneys' glomeruli appear just under the surface of the lining. In an inflamed bladder, these blood vessels may bleed during stretching.

Hunner's Ulcer. These tiny ulcers may appear in the small percentage of IC patients who have the ulcerative form of the disorder.

Small Bladder Capacity. Although bladder capacity can be an issue in several urinary diseases, decreased capacity is one of the hallmarks of IC.

Doctors use several methods to diagnose IC and to look for glomerulations and Hunner's ulcers:

Urine Culture. The doctor examines the urine for red and white blood cells and bacteria that would indicate an infection. If the urine contains no bacteria, IC may be a likely cause of the problem.

Prostate Secretions Culture. To rule out an infection of the prostate gland, the doctor may check for bacteria in a man's prostatic fluid.

Cystoscopy with Hydrodistention. When IC is suspected, this is the preferred method for diagnosis. The doctor will examine the inside of the bladder with a thin, flexible tube called a cystoscope. After the exam, the doctor stretches the bladder to its capacity by filling it with either fluid or gas. This is done while the patient is under general or regional anesthesia. During the cystoscopy and hydrodistention, the doctor will look for bladder wall inflammation, a stiff bladder wall, pinpoint bleeding on the bladder wall, and Hunner's ulcers. He or she will also try to rule out other disorders by spotting bladder stones, tumors, or other abnormalities that could be causing the problem. The doctor may also test how much liquid the bladder can hold while the patient is under anesthesia.

Treatment Options

There is no cure for IC, and treatment is usually a matter of trial and error—trying one or a combination of different therapies to find one that works for the patient. Many people find that a particular treatment relieves their symptoms temporarily, but that their condition returns after a period of weeks, months, or even years.

MEDICATION

This is usually the first line of attack, because it is the least invasive and often most effective form of treatment. Doctors usually try oral medications first. The only oral medication that is FDA-approved to treat IC is pentosan polysulfate sodium (Elmiron). No one knows why it works, but doctors think that it repairs a defect in the bladder lining that makes the bladder susceptible to inflammation. This drug is believed to help around 40 percent of IC patients, but it may affect liver function and therefore its use is carefully monitored by a physician. Gastrointestinal discomfort is another potential side effect. Other oral medications, though not approved specifically for IC, have been successful in alleviating its symptoms. These include aspirin and ibuprofen (to relieve discomfort), anti-inflammatory agents, antihistamines, antispasmodics, anticholinergics (Detrol, Ditropan), muscle relaxants, steroids, tricyclic antidepressants (which may reduce pain and help patients deal with the emotional implications of their problem), and narcotic pain medications (Tylenol with codeine).

If medications don't work when taken orally, doctors may insert them directly into the bladder via a procedure called bladder instillation (or bladder washing). A small catheter is passed through the urethra, and the medicine is injected through the catheter into the bladder. After about 10–20 minutes, the patient urinates to remove the medicine. Treatments are given every week or two for about six to eight weeks, and the cycle can be repeated as necessary. The medication most commonly used, and the only one approved by the FDA, is dimethyl sulfoxide (DMSO). Doctors think this medication works by reducing inflammation and preventing the bladder muscle contractions that cause pain, frequency, and urgency. The only side effect of DMSO is that it leaves a strong garlic-like odor on the breath and skin that can last for up to seventy-two hours after treatment.

DMSO may be used alone or in combination with other medications (such as steroids; heparin, a blood thinner; or lidocaine, an anesthetic) in a drug cocktail. Other medications being used experimentally for bladder instillation are bacillus Calmette-Guerin (BCG), a version of the vaccine used against tuberculosis, and hyaluronic acid (Cystistat).

BLADDER DISTENTION

The same method used to diagnose IC can also be used to treat a percentage of patients. Doctors fill the bladder with water and leave it distended for about eight minutes, then drain it and repeat the process. Bladder distention can be done under general or regional anesthesia. The procedure appears to work by interfering with the transmission of pain signals from nerves in the bladder. It also may stimulate the production of mucin, a natural protective coating that normally lines the bladder surface. Some patients report that bladder distention relieves their symptoms for up to six months. When the effects wear off, the procedure can be repeated.

TRANSCUTANEOUS ELECTRICAL NERVE STIMULATION (TENS)

Wires placed on the lower back or just above the pubic area transmit mild electric pulses to the nerves near the bladder. The device may be worn for a few minutes to a few hours, two or more times a day. Scientists don't know how the device works, but it may increase blood flow to the bladder, strengthen the pelvic muscles that control the bladder, or trigger the release of substances that block pain. Many patients find relief within three to four months of beginning TENS.

SACRAL NERVE ROOT STIMULATION

This procedure is the surgical version of TENS. Doctors implant electrodes inside the body, which are attached to a unit similar to a pacemaker. The device transmits electrical pulses to the sacral nerves, located at the base of the spinal cord in the lower back. The sacral nerves control the bladder as well as the bowel and pelvic organs. By stimulating the nerves of the bladder, the machine helps patients gain more control over their bladder function. Sacral nerve root stimulation is not yet FDA-approved for treating IC, but it is approved for treating urinary urge and frequency, two symptoms of IC.

BEHAVIOR MODIFICATION

Very often, changes to diet and lifestyle habits are enough to alleviate some of the symptoms of IC. Although no scientific evidence exists to back up the claim, many doctors believe that certain foods irritate the bladder. Avoiding things like alcohol, tomatoes, spices, chocolate, caffeinated and citrus drinks, acidic foods, and artificial sweeteners may reduce pain and inflammation. Many patients also find that smoking worsens their symptoms, and they are advised by their doctor to quit. Gentle stretching, pelvic floor relaxation exercises, and biofeedback may relieve symptoms for many patients. Others try bladder training to teach their bladder to hold more urine and stretch it back to its pre-IC size. Training involves emptying the bladder at specified times and steadily increasing the time between urination (more on bladder training in Chapter 8).

SURGERY

If all other methods have failed, surgery may be used as a last resort. Surgery is invasive and expensive, and it only benefits a tiny percentage of IC patients. The four most commonly used methods are fulguration (burning off Hunner's ulcers using an electrical current or laser beam), resection (cutting and removing the ulcers via special instruments inserted up the bladder through a cystoscope), augmentation cystoplasty (removing the damaged sections of the bladder and replacing them with a piece of large intestine), and cystectomy (bladder removal). These procedures require long recovery times, and patients often report that their symptoms return months later.

Future Directions in Interstitial Cystitis Research

So much about IC remains a mystery that researchers are concentrating on understanding what causes the condition and on developing more effective ways of treating it. They are also looking for better methods of classifying different varieties of IC to allow for easier diagnosis and treatment. With regard to diagnosis, researchers are focusing on urinary markers: substances in the urine of IC patients that are not found in the urine of healthy people. Two markers currently under investigation are GP51 and antiproliferative factor (APF). One day, these less invasive and more accurate markers may replace cystoscopy as the preferred diagnosis technique.

CONGENITAL BLADDER ABNORMALITIES

Doctors don't know what causes the following abnormalities of the bladder, which originate in the developing fetus.

Exstrophy

This extremely rare condition occurs in only 1 in 30,000 births, according to the March of Dimes (2001). In babies born with exstrophy, the bladder forms inside out and protrudes from an opening in the abdomen. Normally, when the fetal bladder develops, skin, muscle, and connective tissue fold over to enclose and seal the organ. Exstrophy occurs when the folding process fails to complete, so the inside of the bladder is left open. Babies born with this condition may also be missing a urethra or sphincter, and may have abnormally rotated or spaced pelvic bones and/or abnormal genitalia. If left untreated, a patient with exstrophy will be unable to control urination and will be more prone to chronic infections and one form of bladder cancer. To correct the problem and give the patient control over urination, doctors reconstruct the bladder and urethra, often through a series of surgeries over a number of years. The first surgery is usually performed forty-eight hours after birth to close the bladder and place it inside the abdomen, seal the abdomen, and put the pelvic bones into proper alignment.

Espispadias

This condition is similar to exstrophy (and actually may occur along with exstrophy) with the exception that the bladder is not inside out, but abnormally small. Children with this condition may have differing degrees of bladder control, depending on the extent of the abnormality. Espispadias also affects the genitalia, often producing a short, flat penis in boys and a split clitoris and abnormally situated urinary opening in girls. Both the bladder and genitals can be repaired surgically, but doctors may need to undertake a number of surgeries to give the child full bladder control and normal-looking genitals. The first operation, which is usually done in the first few days of life, fixes bladder abnormalities. Doctors may wait up to a year or two before trying to repair a child's genitalia.

Congenital Diverticula

Like a balloon squeezing through the openings in a closed fist, a bladder afflicted with congenital diverticula squeezes through a weakness in the bladder wall. There may be many diverticula in one bladder. If the problem occurs at the point where the ureter and bladder join, it may force urine to push back up into the ureter and kidneys, causing an infection. Diverticula may be hard to spot, because they often cause no symptoms (aside from infection), and they are often discovered only during an x-ray or cystoscopy to find the cause of recurrent infections. Surgery is the preferred method to correct the problem.

OTHER BLADDER DISORDERS

Neurogenic Bladder

In healthy individuals, the process of urination is controlled by sensory nerves, which send a message to the brain that the bladder is full, and motor nerves, which trigger the sphincter muscles in the bladder to relax and release the urine. These signals are disrupted in patients with a neurogenic bladder. Either the muscles of the bladder don't get the message to contract and release urine (called underactive), or they receive too many messages and contract too frequently, uncontrollably releasing urine (called overactive). In patients with the underactive form of neurogenic bladder, urine may build up in the bladder to the point where it begins to reflux (back up) into the ureters and kidneys. Urine left stagnating in the ureters and kidneys can harbor the bacteria that cause infections. In the overactive form of the condition, the constant release of urine reduces bladder capacity and forces patients to make a constant dash to the bathroom.

The most common cause of an underactive bladder in children is birth defects of the spinal cord, for example spina bifida. An overactive bladder is usually caused by an injury or disorder, such as multiple sclerosis, that interrupts messages sent to and from the brain and spinal cord and the bladder.

To diagnose the problem, doctors use an electroencephalography (EEG) to measure changes in the patient's brain waves and determine if the problem is being caused by a brain dysfunction. Doctors may also test bladder capacity via **cystometry**—filling and emptying the bladder with fluid via a catheter to see how much it can hold and determine whether it is fully emptying.

Treatment depends on the root of the problem and the type of urinary dysfunction it is causing. Patients whose bladders are unable to release urine properly may benefit from intermittent or continuous catheterization to empty the bladder and prevent it from becoming overstretched. If the bladder muscles are overactive and contracting too frequently, medications that relax the bladder, such as propantheline (Pro-Banthine) and oxybutynin (Ditropan) may be of help. If the bladder is underactive, doctors may prescribe drugs such as bethanechol (Urecholine), which simulate the parasympathetic nerves (these nerves direct the urination process by releasing the neurotransmitter acetylcholine, which causes the detrusor muscle in the wall of the bladder to contract). Long-term management of neurogenic bladder involves fluid regulation and bladder retraining (see Chapter 8).

Fistula (Hole in Bladder)

A fistula can be any type of abnormal opening that occurs either between two organs or between an organ and the outside of the body. Fistulas usually follow complications during childbirth (i.e., prolonged labor or an unusually large baby), pelvic surgery (hysterectomy), or as the result of an injury. In women, the opening may occur between the bladder and the vagina (vesicovaginal fistula) or between the urethra and the vagina (urethrovaginal fistula). Aside from persistent urine leakage, patients may have pain in the affected area or an abnormal vaginal discharge. To diagnose the problem, doctors usually fill the bladder with liquid via a catheter to look for any leakage of urine out of the vagina. The only treatment is to surgically repair the opening.

Urethral Stenosis

After menopause, many women experience an increased urinary urgency or frequency, or a need to urinate excessively at night (nocturia). This is caused by a narrowing or tightening of the urethra, called urethral stenosis. During menopause, the ovaries slow their production of the hormone estrogen, which normally keeps the vagina and urethra firm. Without estrogen, the vagina begins to atrophy and can often squeeze down on the urethra, partially closing the passageway through which urine leaves the body. The bladder must therefore work harder to expel urine, and that extra effort tightens its muscles and reduces its ability to stretch to accommodate large volumes of urine. Normally, a healthy woman urinates with about 30 centimeters of pressure. Compare that to a woman whose bladder has become overworked by chronic urethral stenosis—her bladder pressure will likely measure 40–60 centimeters while her bladder is at rest. The increased exertion may also make the bladder feel empty before it really is. The reduced capacity, coupled with residual urine in the bladder, may force women with this condition to run to the bathroom constantly.

To rule out other conditions with similar symptoms, doctors will usually diagnose urethral stenosis by first conducting a complete urologic evaluation, including a bladder cystoscopy to look for stones, tumors, or other abnormalities. The next step is to test bladder capacity and measure the strength of contractions as the bladder fills and empties.

Two methods are commonly used to treat patients with urethral stenosis: estrogen replacement therapy (ERT) and a procedure called dilation. The long-term use of estrogen therapy is widely debated in the medical community because it has been associated with an increased risk of breast cancer as well as other types of cancers, gallbladder disease, blood-clotting disease, and hypertension. But ERT has been shown to help counteract vaginal atrophy and soften the urethra, often restoring normal urine flow. Doctors can also stretch the urethra by inserting progressively larger probes until it is dilated to its full opening. Dilation teaches the bladder to relax and eases the pressure to constantly urinate.

Cystocele (Fallen Bladder)

Cystocele (see Figures 6.2 and 6.3), or fallen bladder (also called prolapse), occurs when the wall between a woman's bladder and vagina weakens, allowing the bladder to droop down into the vagina. The bladder may stretch into the opening of the urethra, causing urine to leak when any pressure is put on the bladder (from a cough, sneeze, or laugh). Cystocele may result from the strain of giving birth, repeated heavy lifting, or persistent straining during bowel movements. Like urethral stenosis, menopause may also be a culprit, because estrogen normally maintains muscle strength around the vagina. There are three grades of cystocele, ranging from the mildest to the most severe. In grade 1 cystoceles, the bladder drops only partway into the vagina. Grade 2 cystoceles reach the vaginal opening, and by grade 3, the bladder protrudes through the vaginal opening. A doctor will be able to see grade 2 and 3 cystoceles upon physical inspection of a woman's vagina, but diagnosis of grade 1 relies on x-rays of the bladder taken during urination.

Mild cystoceles often need no medical attention, but more serious cases

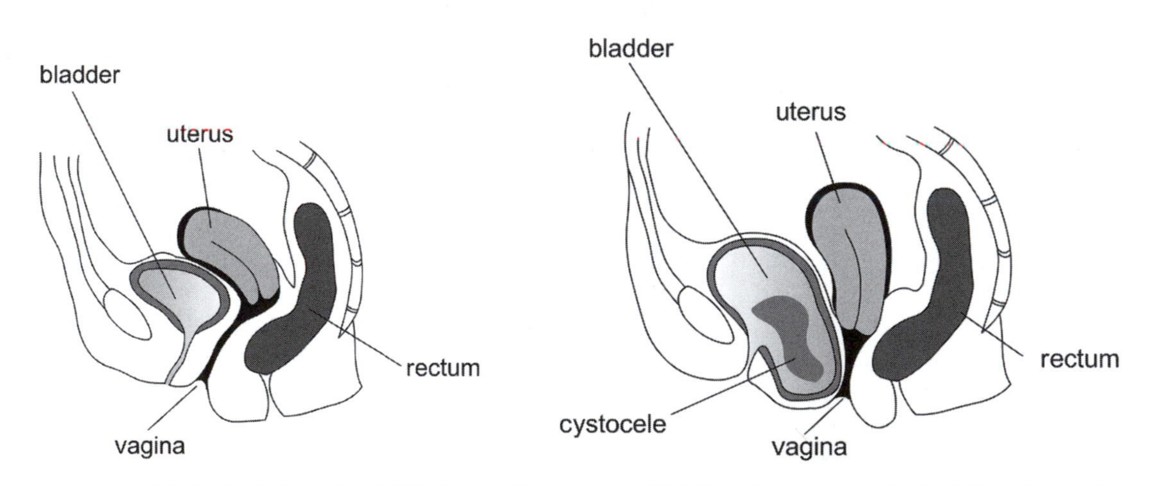

Figures 6.2 and 6.3. Normal bladder position and cystocele.
In a healthy woman, the bladder is raised and supported. In the case of a cystocele, the bladder droops down into the vagina.

may require surgery to reposition the bladder. Doctors may also use a pessary, a device placed in the vagina to support the bladder, or ERT to strengthen the muscles around the vagina and bladder in postmenopausal women.

Vesicoureteral Reflux (VUR)

Normally, the ureterovesical valve at the junction of the ureter and bladder allows urine to flow into the bladder while preventing it from flowing back up toward the kidney. But in patients with **vesicoureteral reflux (VUR),** this valve malfunctions, and urine collected in the bladder pushes back up the ureter and into the kidney, where it can stagnate and cause infection. The condition is usually diagnosed at an early age in children who constantly visit the pediatrician complaining of urinary tract infection symptoms (pain or burning during urination, a frequent need to urinate, or blood or pus in the urine).

There are two types of VUR. In primary VUR, a child is born with a faulty valve at the intersection of the ureter and bladder. Alternately, the ureter may not have grown long enough during fetal development, or it may have grown at the wrong angle to snugly fit into the bladder opening. Primary VUR usually improves as the ureter grows. Secondary VUR is caused by a blockage somewhere in the urinary system. The blockage may be triggered by a bladder infection that causes the ureter to swell.

Doctors diagnose VUR by first looking for signs of a urinary tract infection (blood or bacteria) in a urine sample. The urine test may be followed by images taken of the bladder and urinary system (cystourogram) to determine whether a defect is responsible for the problem. If a urinary tract infection is to blame, antibiotics can usually cure the problem. Primary VUR sometimes requires surgery to fix a faulty valve or to correct the angle at which the ureters connect with the bladder.

Benign and Cancerous Prostate Growth

The primary function of the walnut-sized prostate gland (see Figure 7.1) is to add liquid to the man's sperm, providing nourishment that will sustain the sperm on its journey through the woman's fallopian tubes. The prostate is actually part of the reproductive system (which is covered in its own volume of this series), not the urinary system. But the gland's proximity to the bladder and urethra assigns it a crucial role in a man's urinary health.

The prostate is located at the bottom of the bladder, surrounding the top of the urethra (the tube that transports urine out of the body) like a doughnut. When a man reaches middle age, the prostate becomes prone to growth or inflammation. Any enlargement of the prostate, whether benign or cancerous, can constrict the urethra and inhibit the flow of urine. Imagine stepping on a garden hose and watching the pressure within the hose rise as the water has nowhere to go but back. The effect is similar when the prostate clamps down on the urethra. Urine is forced back up into the bladder, where it can stagnate and cause an infection. Prostate enlargement can make urination difficult and often painful, and can eventually lead to kidney problems, repeated urinary tract infections, bladder infections, and urinary retention.

BENIGN PROSTATE ENLARGEMENT

At birth, the prostate is about the size of a pea. During puberty, the prostate begins to grow rapidly until it reaches its full size. At that point,

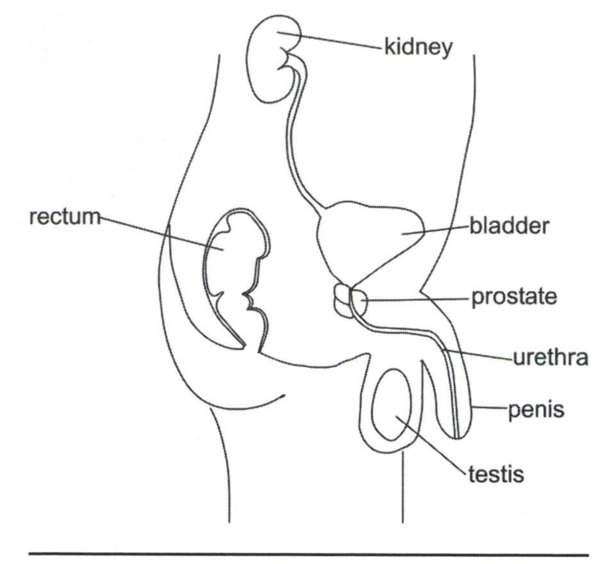

Figure 7.1. The prostate gland.
The prostate is located just under the man's bladder, encircling the top of the urethra.

growth ceases. But for some reason, factors relating to middle age trigger a period of regrowth starting around age 40 (see Figure 7.2). This growth is known as benign prostatic hyperplasia (BPH) or benign prostatic hypertrophy. Unlike prostate cancer, BPH cells are benign—they grow in an orderly fashion and never spread beyond the prostate. Having BPH does not mean that a man has prostate cancer or even that he is predisposed to cancer, although both conditions may simultaneously coexist.

About 80 percent of men will develop BPH by the time they reach their 70s or 80s, according to the American Foundation for Urologic Disease ("Prostate Disease," 2002). The good news is that the condition is not life-threatening, but it can be a nuisance. As the prostate grows, it starts to clamp down on the urethra. Because the bladder must contract more forcefully to expel urine, the muscles in its walls stretch and lose their elasticity, eventually weakening to the point where they can no longer empty the bladder fully. When urine that cannot be expelled begins to stagnate, the result is often one urinary tract infection after another (see Chapter 9), or chronic bladder stones. As the condition worsens, a small percentage of men become unable to urinate, and a few may suffer permanent bladder or kidney damage.

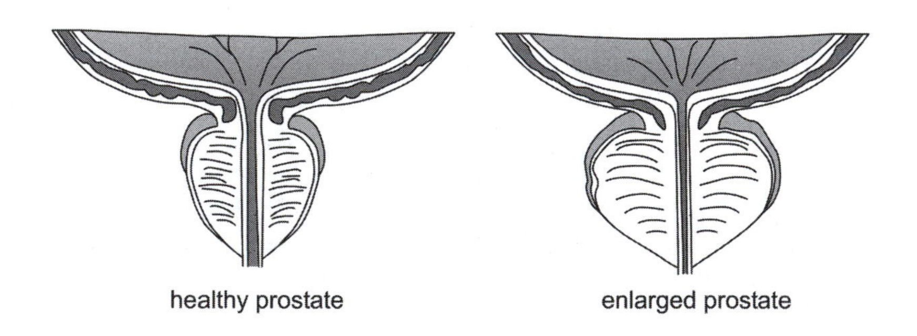

Figure 7.2. A healthy vs. an enlarged prostate.
When the prostate is normal size, it places no pressure on the urethra. But when the prostate grows due to BPH, it closes off the urethral opening, blocking the path of urine.

What Causes BPH?

Doctors are still unsure what mechanisms trigger BPH, but they do have a few theories. One has to do with the male hormone testosterone. As a man ages, the level of testosterone in his bloodstream diminishes, leaving him with higher levels of the female hormone estrogen. Animal studies suggest that increased estrogen in the prostate promotes the cell growth that leads to BPH. A second theory centers on a testosterone derivative called dihydrotestosterone (DHT), which continues to be produced in the prostate even after a man's testosterone level diminishes. DHT may also play a role in cell growth.

Symptoms of BPH

Prostate enlargement itself carries no symptoms, but as the prostate puts pressure on the urethra, the resulting blockage can interfere with urination. Indications of a blockage include:

- Difficulty starting urination
- Weak or intermittent stream
- Consistent feeling of urgency
- Leaking or dribbling of urine
- More frequent urination, especially at night (nocturia)
- Urinary tract infections

Some men are not aware that they have BPH until they are suddenly unable to urinate at all, a potentially dangerous condition known as acute urinary retention.

Testing for BPH

Routine testing for BPH and prostate cancer begins when a man is in his 40s and continues annually for the rest of his life. The examination for BPH begins with a medical history and physical exam, followed by a digital rectal exam (DRE), urine test, and prostate-specific antigen (PSA) blood test.

Digital Rectal Exam (DRE). The doctor inserts a gloved, lubricated finger (see Figure 7.3) into the man's rectum to feel the prostate for signs of enlargement or any unusual lumps or hard spots (which could be cancerous).

Urine Test. A urine sample is analyzed in the lab to rule out other urinary tract conditions with symptoms similar to BPH, such as a urinary tract infection, prostatitis (see the discussion on prostatitis later in this chapter) cystitis, or kidney disease.

PSA Blood Test. Prostate-Specific antigen (PSA) is a protein produced by the prostate gland. Normally, it circulates in low levels in a man's bloodstream.

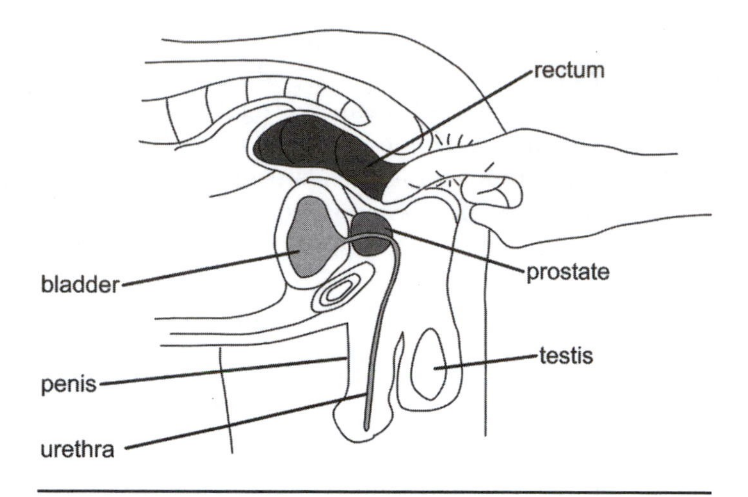

Figure 7.3. Digital rectal exam.
To check for benign or malignant prostate growth, the doctor inserts a gloved, lubricated finger up the patient's rectum and feels for any unusual lumps or hard spots.

But when the prostate is enlarged or inflamed, it releases excess PSA into the blood. Because elevated PSA levels may indicate any one of three conditions—BPH, prostatitis, or prostate cancer—a high reading usually warrants further testing (see the section on diagnosing prostate cancer later in this chapter).

Depending on the results of the tests listed above, the doctor may call for one of the following urinary tract studies:

Urodynamic Evaluation. A series of tests that determine the type of voiding dysfunction by gauging how well a patient's bladder, urethra, and rectum are holding and releasing urine. Using a variety of techniques, the doctor measures bladder pressure and analyzes urine flow during bladder filling and emptying. Tests include:

Uroflometry. The patient urinates into a special device that electronically calculates the flow rate and the amount of urine voided. Flow rates greater than 15 ml/sec. are considered normal. Anything under 10 ml/sec. is a sign of obstruction.
Cystometry. The patient's bladder is filled with liquid via a **catheter**, and a device called a cystometer electronically measures bladder pressure during filling and emptying.

Postvoid Residual Measurement. After the patient has urinated, the doctor inserts a catheter into the bladder, through which any remaining urine is drained and measured. A volume of less than 50 milliliters postvoid is considered normal; more than 200 milliliters can indicate a bladder emptying problem.

Ultrasound. Using the same sound wave technology that allows a pregnant woman to view her unborn baby in the womb, doctors can measure

the prostate for BPH growth and look for cancerous tumors. A probe placed in the man's rectum directs sound waves at the prostate, which form an image of the gland on a video screen.
Cystoscopy. A thin lighted instrument called a cystoscope is threaded through the opening of the penis and into the bladder. The device allows the doctor to see if the prostate is squeezing the urethra and to look for signs of bladder cancer or stones.
Intravenous Pyelogram (IVP). Dye is injected into the patient's vein. When the pelvic area is x-rayed, the dye reveals any urinary tract obstructions.

Treatment Options

Not all patients require treatment for BPH. Men who are not in any discomfort can opt for "watchful waiting," a program of careful monitoring that involves a yearly visit to the urologist for a DRE and blood and urine tests. But if symptoms worsen, or if the condition progresses and the patient suffers kidney damage, prostate bleeding, severe or recurrent urinary tract infections, or bladder stones, one of the following treatments is necessary.

MEDICATION

- *Alpha Adrenergic Blockers.* Originally designed to treat high blood pressure, these drugs relax the smooth muscles of the prostate and bladder neck to relieve the obstruction and allow urine to once again flow freely. Currently, the most commonly prescribed alpha adrenergic blockers are terazocin (Hytrin), doxazosin (Cardura), and tamsulosin hydrochloride (Flomax). These drugs work quickly, alleviating symptoms within a couple of days, but they do have a few side effects. Because they lower blood pressure, alpha adrenergic blockers can cause lightheadedness, dizziness, fatigue, or headaches, especially in men who already suffer from low blood pressure.
- *5 Alpha-reductase Inhibitors.* These drugs shrink the prostate by blocking conversion of testosterone to DHT, which plays a role in prostate enlargement. 5 alpha-reductase inhibitors can relieve the urinary blockage and eliminate the need for surgery, but they may take anywhere from six months to a year to be fully effective. Finasteride (Proscar) is currently the only 5 alpha-reductase inhibitor approved by the U.S. Food and Drug Administration (FDA). Side effects can include decreased libido, erectile dysfunction (impotence), and reduced ejaculate volume. Also, men who are taking this medication must be more carefully screened for prostate cancer, because the medication can influence blood PSA levels.

SURGERY

Reserved for men with the most severe symptoms (kidney damage, frequent urinary tract infections, incontinence, or bladder stones), surgery is considered the most effective BPH treatment, but it also comes with the greatest risks. The following surgical procedures work by removing the core of prostate tissue, which relieves pressure on the urethral opening through which urine flows from the bladder:

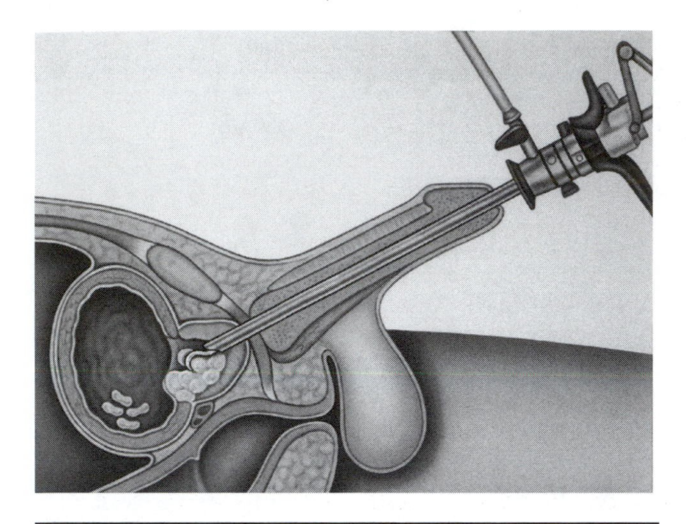

An illustration depicting transurethral (TURP) resection of the prostate gland, in which a resectoscope is inserted into the penis, which removes benign or cancerous tissue from the prostate. © R. Lazarus/Custom Medical Stock Photo.

- *Transurethral Resection of the Prostate (TURP).* Doctors consider TURP (see illustration) the most effective BPH remedy, and it is also the most widely used. This is a surgical procedure; however, no incisions are made. The doctor places a long, thin instrument called a **resectoscope** through the hole in the penis, past the urethra to the prostate. At the end of the resectoscope is either a wire loop or knife, which scoops out tissue from the inside of the prostate. Tissue is retrieved from the bladder with a suction device. When the tissue is removed, it is sent to a pathologist for cancer screening. TURP is extremely effective, especially in the short term, but about a quarter of men will require a repeat of the procedure within ten years. Many men experience impotence and/or incontinence for up to a year after surgery. Another side effect is retrograde ejaculation, in which semen spurts backward into the bladder during ejaculation, rather than out through the penis. Men who have this problem can still feel the pleasure of an orgasm, but are unable to father children.
- *Transurethral Incision of the Prostate (TUIP).* When prostate enlargement is minor, doctors opt for TUIP, in which two small cuts are made in the prostate to increase the urethral opening. This procedure is less invasive than TURP, but its long-term effects are still unknown.
- *Prostatectomy.* Used only when the obstruction is very large, prostatectomy involves the removal of prostate tissue via an open incision. In cases of BPH, only part of the prostate is removed, as opposed to prostatectomy for prostate cancer, in which the entire prostate is often taken out. The surgeon makes an incision in either the patient's lower abdomen or in the **perineum** (the area of tissue that lies between a man's scrotum and anus) and removes all of the prostate tissue causing the blockage.

HEAT TREATMENTS

Using microwave, radio-frequency, or laser energy, doctors can kill off prostate tissue with heat as effectively as they can remove it with a knife, but with less blood loss and shorter recovery time. Some of the procedures currently in use include:

- *Transurethral Microwave Thermotherapy (TUMT).* Two devices, the Prostatron and the Targis System, use computer-regulated microwave energy to heat and destroy prostate tissue. The Prostatron emits microwave energy through a catheter inserted up the patient's rectum. Cool water circulates around the tip and sides of the catheter during the procedure to protect the urethra from heat damage. The Targis System is similar to the Prostatron in that it uses microwave heat to destroy obstructing portions of the prostate. Both procedures take about an hour to perform and can be done on an outpatient basis without anesthesia. TUMT has not been associated with incontinence or impotence, but its long-term effectiveness remains unknown.

- *Transurethral Needle Ablation (TUNA).* Radio-frequency energy is applied to the prostate via tiny needles inserted through a catheter. The prostate is heated to about 120°F, and the heat burns away the obstructing tissue. TUNA improves urine flow and relieves BPH symptoms with fewer side effects than TURP, but the procedure does not work on men who have very large prostates.
- *Transurethral Electrovaporization of Prostate (TVP).* This new version of TURP uses a special metal instrument that emits a high-frequency electrical current to cut and remove excess prostate tissue. TVP results in less bleeding and shorter hospital stays than TURP, but the long-term effects are still under investigation.
- *Transurethral Evaporation of Prostate (TUEP).* This procedure is similar to TVP, with the exception that it uses laser energy, rather than electrical current, to destroy prostate tissue.

OTHER TECHNIQUES

- *Prostatic Stent.* A tiny wire tube inserted near the prostate pushes back obstructing tissue to widen the urethra. Over time, the tissue will begin to grow over the stent to hold it in place. As with any foreign object implanted in the body, infection and excess bleeding are common side effects.
- *Balloon Dilation.* The doctor pushes a small deflated balloon up through a catheter into the patient's urethra. When the balloon is inflated, it widens the urethra and pushes back the prostate tissue. This procedure is not very effective and is therefore rarely used.

Patients will often experience pain or discomfort during urination for a few months after BPH surgery. Other side effects include incontinence, blood in the urine, and retrograde ejaculation, but these symptoms usually diminish within a year. It is also not uncommon for remaining prostate tissue to trigger cell regrowth, so patients must be carefully monitored after BPH surgery for a recurrence.

New Directions in BPH Research

Doctors believe that the herb saw palmetto may be effective in preventing BPH, because studies show that it improves urinary symptoms associated with the condition. These symptoms include nocturia (urinating at night), weak stream, frequent urination, and residual urine volume (urine left in the bladder from incomplete emptying). It is thought to work because it prevents the prostate from producing the hormone dihydrotestosterone (DHT), which is associated with BPH cell growth. In 1998, researchers at the Department of Veterans Affairs looked at more than a dozen studies involving saw palmetto and concluded it to be equally as effective as the drug finasteride, but without the side effects. Whether saw palmetto will come into practical use for the prevention and treatment of BPH will depend on the results of future studies.

PROSTATE CANCER

A majority of American men will be affected by prostate cancer at some point in their lives. Prostate cancer is one of the leading cancers among men, second only to skin cancer. According to the National Institute of Diabetes and Digestive and Kidney Diseases, more than 168,000 men will develop prostate cancer each year, and more than 30,000 will die (2001). The good news is that the disease strikes so late in life that men are far more likely to die *with* prostate cancer than *from* prostate cancer.

Risk Factors

Prostate cancer is still not well understood, but doctors have identified a number of risk factors:

Age. In the United States, most prostate cancer patients are over age 65. According to the National Cancer Institute, most deaths occur among men aged 70 or older.

Race. African American men face the highest prostate cancer risk in the world, yet no one knows why. Black men are also more likely to develop the disease at an earlier age, and die from it, than any other race. The lowest risk is among Asian and Native American men. Scientists have speculated that Asian men face less of a risk because their diet is low in animal fat (see the diet section that follows).

Family History. A small percentage of prostate cancers are hereditary. Studies have shown that men with a family history (father or brother) of prostate cancer face twice the risk of other men. Scientists have isolated several genetic mutations associated with prostate cancer in families that have a history of the disease. A gene called macrophage scavenger receptor-1 (MSRI), already known for its role in coronary artery disease, was recently implicated in prostate cancer susceptibility among men of African and European descent (Xu et al., 2002). Researchers at Wake Forest and Johns Hopkins Universities believe that mutations in this gene inhibit the ability of macrophages—immune system cells—to clean up after prostate infections, producing an inflammatory response linked to the development of prostate cancer.

Diet. Some studies have found that diets high in fat and animal meat increase a man's risk of developing prostate cancer. Conversely, diets high in fruits and vegetables appear to offer some protection against the disease. Scientists believe that fats increase production of the hormone testosterone, which speeds the growth of cancerous cells.

Symptoms of Prostate Cancer

Many men live with prostate cancer for years without a single symptom, which is why about 40 percent of prostate cancers are not detected until they have begun to spread. In most cases, cancer is discovered not by its symptoms but by a routine medical evaluation or surgical procedure to remove excess prostate tissue.

Once the cancer has begun to spread, a number of symptoms may appear. These include:

- Persistent, dull pain in the lower back, hips, or upper thighs
- Sudden or frequent need to urinate
- Difficulty starting the stream of urine
- Weak or intermittent flow
- Pain or burning during urination
- Blood in the urine (hematuria) or semen
- Difficulty having or maintaining an erection
- Painful ejaculation
- Loss of appetite

Because many of these symptoms are similar to BPH and the inflammatory condition prostatitis, a careful examination by a urologist is necessary to confirm a cancer diagnosis.

Diagnosing Prostate Cancer

The doctor will always start with a careful evaluation of the patient's symptoms, family and personal medical history, and overall health. If prostate cancer is suspected, the two most commonly used diagnostic tools are the digital rectal exam (DRE) and the prostate-specific antigen (PSA) test.

DIGITAL RECTAL EXAM (DRE)

The doctor inserts a gloved, lubricated finger into the man's rectum, feeling for lumps or abnormalities on the back wall of the prostate. This technique is effective at spotting early-stage tumors in the outer portion of the gland, but it may miss up to a third of cancers, which are located deeper inside the prostate.

PROSTATE-SPECIFIC ANTIGEN (PSA) TEST

The patient's blood is drawn and analyzed for the PSA protein (see the section on testing for BPH earlier in this chapter). When the prostate is

The Controversy Over PSA Testing

To PSA test . . . or not to PSA test? That is the question at the center of a fierce debate among medical professionals. On one side of the issue are doctors who argue that the prostate-specific antigen (PSA) test is inaccurate. Anywhere from 12 to 15 percent of men over age 50 will test high for PSA levels yet not have prostate cancer. Many of those men will undergo unnecessary biopsies and/or treatments for a condition that is actually benign (benign prostatic hypertrophy or prostatitis can elevate PSA levels, for example).

On the opposing side of the issue are the doctors who argue that, despite its limitations, the PSA test remains the best prostate cancer screening tool at their disposal, having dramatically increased prostate cancer detection rates since it was introduced in the mid-1980s. Early detection, they say, is crucial to catching prostate cancer before it spreads throughout the body.

Doctors are hopeful that newer, more sensitive PSA screening methods, such as testing free versus bound PSA levels, will lessen the margin of error and make prostate cancer screening a much more precise science in the future.

---

healthy, PSA levels in the blood are low, but any inflammation of the prostate (including BPH, cancer, or prostatitis) will reveal itself with a high PSA reading. A reading of between 4 and 10 nanograms per milliliter (ng/ml) is considered normal, but anything over 10 ng/ml is suspicious. The PSA test is adept at identifying prostate inflammation, but it cannot distinguish between benign and cancerous growth, which leads to a high number of false positive results.

Since the PSA test was first introduced in 1986, it has stirred up fierce controversy within the medical community. (See "The Controversy over PSA Testing.") Some doctors say that it has dramatically improved prostate cancer detection rates and reduced mortalities. Opponents argue that the test has resulted in thousands of unnecessary biopsies and potentially risky surgical procedures. But until scientists come up with a better tool, a PSA test combined with a DRE remains the only medically approved screening method for prostate cancer. Both the American Urological Association (AUA) and the American Cancer Society recommend that all men have a yearly PSA blood test and DRE starting at age 50 (when life expectancy is greater than ten years). Men with a family history can begin their annual checkups as early as age 40.

Until new diagnostic tools are identified, doctors have been working to fine-tune the PSA test by identifying a number of more specific indicators:

- *PSA Velocity, or How Much PSA Levels Rise Over a Period of Time.* Experts have not come to an agreement over how much of a change should be regarded as suspicious, however, because PSA values can drift upward or downward naturally over time.

- *Age-adjusted PSA.* Assigning different ranges of what is considered normal based on a man's age. For example, a PSA reading below 2.5 ng/ml may be considered normal for men under age 50, while men over 70 can have a normal reading as high as 6.5 ng/ml. Doctors don't know whether this method actually works.
- *PSA Density.* The relationship of PSA to the size of a man's prostate. For example, a high PSA level in a man with a large prostate would be of less concern than the same measurement in a man with a small prostate.
- *Free versus Attached PSA.* PSA circulates in the blood in two forms: free or attached to a protein molecule. Scientists have found that men with prostate cancer have more of the attached form, and men with benign prostate growth have more of the free form. In 1998, the FDA approved the Tandem R test, which measures free PSA levels. Doctors hope that by distinguishing between the two forms of PSA, they'll be able to prevent at least a percentage of unnecessary biopsies in patients with borderline PSA blood readings.

The only way to diagnose prostate cancer with any confidence is by taking a tissue sample and examining it under a microscope (called a biopsy). If the doctor suspects cancer based on the results of the DRE and PSA tests, he or she will remove a sliver of prostate tissue via a narrow needle inserted in the man's rectum. The needle is guided by an ultrasound probe, which uses sound waves to direct the doctor to any suspicious-looking tissue. The tissue is then sent to a pathologist, who views it under a microscope to determine how fast the cancer is growing (the grade) and how far it has spread (the stage) (see Stages of Prostate Cancer). The degree to which tumors have progressed is determined by the Gleason grading system, which assigns a number based on the shape and microscopic appearance of tumor cells. The higher the grade, the further the cancer has progressed.

TRANSRECTAL ULTRASOUND

A small probe is placed inside the rectum, which bounces sound waves off the prostate (see photo). The sound waves create a picture of the prostate, called a sonogram, on a video screen. Doctors look for patterns that distinguish a cancerous tumor from healthy prostate tissue. Some ultrasound machines come with a spring-loaded "gun" that allows the doctor to remove suspicious-looking tissue for biopsy.

BONE SCAN

In about a third of patients, prostate cancer spreads to the bone, where it is often painful and can ultimately be deadly. Doctors detect bone metastases by injecting a radioactive solution into the patient's bloodstream. The

Stages of Prostate Cancer

Stage I. The cancer is too small to be felt during a rectal exam and most likely causes no symptoms. At this stage, tumors are usually discovered during prostate surgery to treat BPH or other prostate growths. These early tumors have not spread beyond the prostate.

Stage II. The tumor is large enough to be felt during a rectal exam but has not spread beyond the prostate.

Stage III. The cancer has spread to tissues surrounding the prostate.

Stage IV. The cancer has spread to the lymph nodes, bone, liver, lungs, or other part of the body. (Regardless of where the cancer resides, it bears the name prostate cancer.)

The Gleason grading system evaluates the extent to which the tumor cells and glands resemble normal prostate cells.

Grade 1. Cancerous cells resemble normal prostate tissue, and the tumor is not expected to grow quickly. Grade 1 is the most well-differentiated and slowest growing prostate cancer.

Grades 2–4. These tumors fall in between grades 1 and 5 in terms of differentiation. The higher the number, the faster the tumor is growing.

Grade 5. Cancerous cells are very irregular, and differ markedly from normal prostate cells. Grade 5 cells are the most poorly differentiated and quickly growing prostate cancer.

radioactive material is drawn to cancerous bone cells, which show up on the scan as "hot spots."

COMPUTERIZED TOMOGRAPHY (CT) SCAN

A rotating x-ray beam creates a three-dimensional, cross-sectional view of the pelvis that reveals whether cancer has reached the lymph nodes. This is an important diagnostic tool, because the lymph nodes are a jumping off point from which cancer can quickly spread throughout the body. Because the CT scan does not reveal very early cancers and is not adept at determining cancer stage, the test is most effective when combined with other screening tests.

MAGNETIC RESONANCE IMAGING (MRI)

Like the CT, the MRI produces a detailed three-dimensional image of the body, but it uses magnetic fields rather than x-rays. An MRI is used to detect whether the cancer has spread to the lymph nodes, seminal vesicles, bladder, or bone.

LYMPH NODE BIOPSY (LYMPHADENECTOMY)

If tests indicate that cancer has reached the lymph nodes, the doctor will remove some of the nodes closest to the prostate for further examination. Nodes are removed in one of three ways: via an incision in the abdomen, via a thin needle inserted directly into the lymph nodes, or through a laparoscope—a long slender tube inserted into the abdomen. A lymphadenectomy is usually used to confirm that cancer is confined to the prostate.

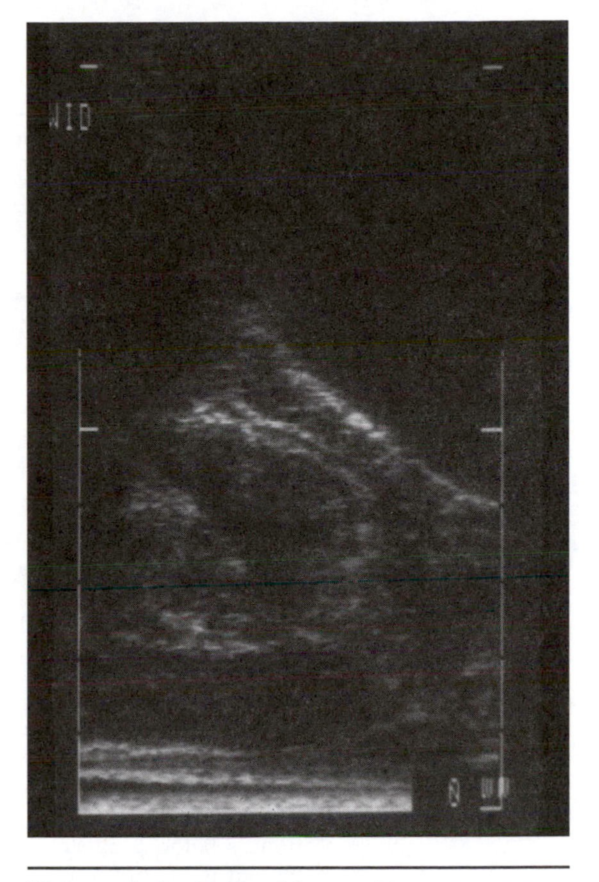

Ultrasound of a 68-year-old man with prostate cancer. © SIU BioMed/Custom Medical Stock Photo.

Treatment Options

Which of the following treatments the doctor chooses will depend on the stage of the cancer and the patient's overall health:

Watchful Waiting. Sometimes, the best treatment for prostate cancer is no treatment at all. When the patient is elderly or in poor health, he may face greater risk from surgery-related complications or drug-related side effects than from the disease itself, especially if the cancer is slow-moving. Of course the word "watchful" is key in this equation, and patients are carefully evaluated with PSA blood tests and DREs every six months.

Surgery. In younger men (under 70) whose cancer has not spread beyond the prostate, removing part or all of the gland in a procedure called a prostatectomy offers the greatest chance for long-term survival (see Survival Rates in Table 7.1). When the prostate is removed completely, the procedure is called a radical prostatectomy. A landmark study (Holmberg et al., 2002) found that, compared with watchful waiting, prostate removal significantly reduced the risk of death from prostate cancer without dramatically affecting quality of life.

A prostatectomy can be performed in one of several ways:

- Radical retropubic prostatectomy (RRP). Doctors remove the prostate, nearby lymph nodes, and seminal vesicles through an abdominal incision. This is the most common type of prostate removal surgery, because the

TABLE 7.1. Prostate Cancer Survival Rates

Years with Prostate Cancer	Survival Rate
5 Years (when diagnosed in local or regional stages)	100 percent
5 Years (when diagnosed at any stage)	96 percent
10 Years (overall survival rate)	75 percent
15 Years (overall survival rate)	54 percent

Source: American Cancer Society, *Cancer Facts & Figures 2002.*

large incision allows doctors to spare the nerves surrounding the prostate that control erection and continence.

- *Radical Perineal Prostatectomy (RPP).* The surgeon makes a small incision in the perineum, through which he or she removes the entire prostate. The lymph nodes may also be removed through a separate incision in the abdomen. This procedure produces less blood loss and requires a shorter hospital stay than RRP, but the smaller incision makes nerve-sparing surgery almost impossible, and impotence is a common side effect.
- *Transurethral Resection of the Prostate (TURP).* The doctor removes part of the prostate with an instrument inserted up through the urethra. TURP is used primarily to open the urethra when BPH enlargement has blocked the urine pathway (see the section on Benign Prostate Enlargement earlier in this chapter), but this procedure can also reduce the size of cancerous tumors before surgery or radiation, or it can be used in cases where the patient is too old or too ill to undergo a radical prostatectomy.
- *Cryosurgery.* An ultrasound-guided probe delivers liquid nitrogen to the prostate. The extreme cold freezes and destroys cancer cells. To protect nearby tissue, a catheter filled with a warming solution is placed inside the urethra. Studies on this technique have so far yielded mixed results, with success rates varying based on the size and stage of the tumor.

Any one of the surgeries mentioned above can result in urinary incontinence, impotence, or rectal damage. Nerve-sparing RPP is somewhat effective in preserving continence and potency, but patients may take up to a year to fully recover from surgery. Even when a man is able to achieve and maintain an erection following surgery, he will not be able to father children. He will still be able to feel pleasure during sexual intimacy, but his orgasms will be dry because his prostate no longer produces semen.

RADIATION

Radiation therapy uses high-powered radioactive waves to kill prostate cancer cells. Radiation is most effective when the cancer is confined to the prostate, but it can also be used once the cancer has spread. Doctors sometimes recommend radiation before or after surgery to shrink the tumor or to remove cancerous cells that surgery might have left behind. When a patient's cancer has progressed beyond the operable stage, radiation can be given to alleviate pain.

There are two kinds of radiation: external beam radiotherapy, and internal radiation or **brachytherapy**.

- *External Beam Radiotherapy.* The most common form of radiation is much like getting an x-ray, except that the x-ray machine is focused over the patient's body for a longer period of time. The large machine delivers a beam of radiation, which destroys prostate cancer cells. Each treatment lasts about fifteen minutes and is repeated five days a week for about six weeks. The major drawback to external beam radiation is that it kills healthy as well as diseased tissue. But a newer technique, called intensity modulated radiation therapy (IMRT), more precisely targets the radiation beam to spare healthy tissue. IMRT creates a computer-generated 3-D image of the prostate to guide the technician directly to cancerous cells. Unlike the single beam of traditional radiation machines, IMRT sends out thousands of tiny "beamlets," which come together at the prostate to deliver a more intense but highly focused radiation dose.
- *Internal Radiation (brachytherapy).* Tiny radioactive seeds, no bigger than grains of rice, are injected into the prostate via ultrasound-guided needles. These seeds carry double the radiation dose of external beams. Because they are placed at the site of the tumor, the tiny radiation carriers can wipe out tumors without destroying healthy tissue. This procedure is still relatively new, but early studies are promising, especially in men whose cancer is confined to the prostate.

Common complaints following radiation therapy include bowel and bladder irritation, diarrhea, rectal leakage, frequent urination, and blood in the urine. Impotence is one of the most serious and prevalent side effects, eventually developing in up to 50 percent of radiation patients. Internal radiation generally produces fewer side effects than external radiation.

HORMONE THERAPY

When the cancer has spread and surgery is not an option, hormone therapy can alleviate the symptoms of prostate cancer, but it cannot cure the disease. Hormone therapy follows the premise that at least some prostate cancer cells depend on male hormones (**androgens**) to survive. If these hormones are suddenly unavailable, the theory is that cancerous cells will die. Depriving the body of hormones is effective at shrinking tumors, but eventually the cancer becomes wily enough to grow without hormones, and the disease once again progresses.

Types of hormone therapy include:

Orchiectomy. The quickest and least expensive way to halt the production of male hormones (specifically testosterone) is by removing their source—the testicles. This surgery, also known as castration, has become less popular with the recent availability of hormone-blocking drugs.

Luteinizing Hormone-releasing Hormone (LHRH) Agonists. An alternative to surgery, drugs like leuprolide (Lupron) and goserelin (Zoladex) decrease testosterone production in the testicles. LHRH agonists are administered monthly or every three to four months by injection.

Adrenal Androgen Inhibitors *and* ***Antiandrogen Agents.*** Most of the androgens in a man's body are produced by his testicles. Cut this hormone supply off, and most—but not all—hormone production ceases. A small amount of male hormones continues to be produced by the adrenal glands (which lie just above the kidneys), which can continue to feed prostate cancer cell growth. Adrenal androgen inhibitors such as ketoconazole (Nizoral) and aminoglutethimide (Cytadren) stop the adrenal glands from producing androgens, and antiandrogen agents such as flutamide (Eulexin) and bicalutamide (Casodex) block the body's ability to use androgens. To completely inhibit hormone production, doctors may prescribe a combination of LHRH agonists and antiandrogen agents (called a total androgen blockade). Whether this combined therapy is more effective than either treatment alone remains to be seen.

The most notable and common side effect of hormone therapy is impotence, which can occur in nearly all patients who undergo orchiectomy and in a majority of patients who take LHRH agonists. The loss of male hormones can also lead to reduced sexual desire, nausea, vomiting, diarrhea, hot flashes, and breast growth or tenderness. A total androgen blockade typically causes more severe symptoms than single hormone therapies.

New Directions in Prostate Cancer Treatment

GENE MARKERS

Researchers are hoping to isolate genetic markers that will help them identify men at risk for prostate cancer. Several genetic mutations have been identified that are believed to increase prostate cancer susceptibility. One of these genes, called ribonuclease L or RNASEL, has been mapped to what is called the Hereditary Prostate Cancer 1 Region of chromosome 1. RNASEL normally protects cells against viruses and plays a role in cell death. Scientists believe that a mutation may interfere with RNASEL's ability to help cells self-destruct, which could explain why prostate cancer cells begin to multiply out of control.

Scientists are also hoping to use genetic markers to develop more sensitive and accurate prostate cancer–screening methods. One such gene, a protein called a-methylacyl-CoA racemase (AMACR), is under investigation as a replacement for the PSA test. Compared with the PSA blood test, which cannot differentiate between prostate cancer and benign cell growth, the AMACR test has been shown in studies to be about 97 percent accurate, with no false positives.

MEDICATION

The testosterone-blocking drug finasteride (Proscar) is currently under investigation as a new hormone therapy for prostate cancer. Other drugs that have shown promise include angiogenesis inhibitors such as thalidomide, which cut off the blood supply to tumors, and immunotherapeutic agents such as interleukin-2, which teach the patient's immune system to seek out and destroy cancerous cells. There is also preliminary evidence that daily use of nonsteroidal anti-inflammatory drugs (NSAIDs) such as aspirin and ibuprofen may offer some protection against prostate cancer, especially in men over 60, but additional research is needed before doctors can begin recommending NSAIDs to their patients.

PSA VACCINE

A vaccine made from a fragment of the PSA protein, the same protein used to detect prostate cancer, has been shown in studies to kill prostate cancer cells. When the vaccine is injected, it causes white blood cells from the patient's immune system to turn into killer cells and selectively destroy the cancerous cells that express PSA. This still-experimental method offers advantages over other therapies in that it can be specifically designed for the patient's immune system and it kills cancerous cells while sparing healthy cells.

DIET

Scientists have recently turned their attention to three antioxidants—selenium, vitamin E, and lycopene—which they believe offer some protection against prostate cancer. Researchers at Stanford University (Brooks et al., 2001) discovered that men with high levels of selenium (a trace element found in foods such as grains, fish, and animal organ meats) in their bloodstream were up to five times less likely to develop prostate cancer than men who had very little of the element. Doctors believe that selenium works by acting as an antioxidant, neutralizing toxins in the body called free radicals, which can damage healthy cells and eventually lead to cancer. A seven-year prostate cancer prevention trial is currently studying the effects of selenium on more than 32,000 American men.

Lycopene, an antioxidant found in tomatoes, may also protect against prostate cancer. Studies have found that men who regularly eat tomato sauce, ketchup, and other tomato-based foods lower their risk of prostate cancer by anywhere from 35 to 50 percent. Tomatoes cooked in oil are said to have the greatest benefit, because cooking breaks down the cell walls to release greater amounts of lycopene.

PROSTATITIS

Prostatitis is really a catchall term to describe a condition doctors know very little about. It refers to any inflammation of the prostate, but in almost all cases, the cause of the swelling is a mystery.

There are three types of prostatitis: acute infectious, chronic infectious, and noninfectious.

Acute Infectious. This is the least common but most severe form of prostatitis. Acute infectious prostatitis occurs when bacteria from the urinary tract or large intestine make their way to the prostate. The infection usually comes on very suddenly and can be severe enough to require hospitalization.

Chronic Infectious. Like acute infectious prostatitis, this form of the condition is brought on by a bacterial infection. But unlike the acute version, it develops very slowly and lingers over time.

Noninfectious. This is by far the most common form of prostatitis, yet it is also the least understood. The prostate becomes inflamed; however, lab tests reveal no evidence of a bacterial infection. Some doctors have theorized that noninfectious prostatitis may in fact be bacterial in nature, caused by a bacteria that has not yet been identified.

Symptoms of Prostatitis

Some patients experience no symptoms with prostatitis, whereas others endure painful and difficult urination, flu-like chills, fever, and body aches (see Table 7.2). When symptoms do occur, they are often difficult to diagnose as prostatitis because they mimic the symptoms of urinary tract infections and other prostate conditions. One condition commonly mistaken for prostatitis is prostatodynia, pain in the pelvic or perineal region caused by abnormal nerves or muscle spasms along the pelvic floor.

Diagnosing Prostatitis

When the doctor suspects prostatitis, he will look for inflammation via a DRE (see the section on testing for BPH earlier in this chapter). If the prostate is swollen, the doctor will collect prostate fluid and urine to determine whether the cause is bacterial or nonbacterial. A commonly used procedure is known as the three-glass urine test. First, the patient urinates into a container, but stops before his bladder is completely empty. The doctor then massages the patient's prostate and collects the fluid that pools in the urethra. Finally, the patient finishes urinating into a third container. All three samples are sent to a lab and tested for the presence of bacteria (which would indicate an infection) and white blood cells (which alone would indicate a nonbacterial inflammation).

Treatment Options

Bacterial prostatitis (acute infectious prostatitis or chronic infectious) is the easiest to treat. Antibiotics usually clear up the infection within one to two weeks. Chronic infectious prostatitis may require a four- to twelve-week dose of antibiotics. Patients with chronic prostatitis that doesn't clear up

TABLE 7.2. Symptoms of Prostatitis

Acute Infectious	• Chills and fever • Body aches • Pain in the prostate gland, lower back, or genitals • Increased urinary frequency or urgency • Difficulty urinating • Pain or burning during urination • Blood in the urine • Painful ejaculation
Chronic Infectious	• Slight fever • Pain in the lower back or perineum • Frequent urge to urinate • Difficulty starting the stream of urine • Weak or intermittent stream • Pain or burning during urination • Blood in the semen or urine • Pain during ejaculation • Recurrent bladder infections
Noninfectious	• Frequent urge to urinate • Difficulty starting the stream of urine • Weak or intermittent stream • Pain or burning during urination • Pain during ejaculation • No bacteria show up in urine and prostate fluid tests

after a few months may require a continuous low dose of antibiotics for a period of several months. In rare cases, doctors may need to surgically remove the infected portion of the prostate if antibiotics prove ineffective. Noninfectious prostatitis is much harder to treat, because doctors don't know what is at the root of the problem. In some cases, patients with nonbacterial prostatitis do respond to low doses of antibiotics, but treatment is more often a matter of trial and error. Patients might be given an alpha blocker to relax their prostate muscles and relieve urination difficulties, or a pain reliever such as ibuprofen to ease fever and muscle aches. Warm tub baths, regular exercise, and dietary changes (avoiding alcohol, coffee, and spicy foods if they exacerbate symptoms) may also be effective in relieving a patient's discomfort.

Urinary Incontinence

Millions of Americans have a problem so embarrassing that they are afraid to discuss it, even with their doctors. This affliction has stripped them of their dignity and robbed them of their freedom. Incontinence, or the loss of control over urination, is both physically uncomfortable and psychologically debilitating. Many who suffer from it rarely leave their homes for fear of wetting in public. Some report diminished quality of life or feelings of depression as a result of social withdrawal and isolation. Frequently, older adults are placed in institutions because family members are unsure of how to deal with the problem.

According to the National Association for Continence, incontinence afflicts about 13 million adults, the majority of whom are women. The problem can range in severity from a few drops of urine lost during a cough or laugh to a sudden urge to urinate. When the urge hits, the patient may have no control over urination and the bladder simply empties. Incontinence is not a disease itself, but is instead a symptom of other conditions. The good news is that a number of treatments can either cure or improve the symptoms of incontinence. The bad news is that many people are reluctant to visit their doctor out of embarrassment or because they harbor the misperception that treatments are expensive and ineffective.

THE URINATION CYCLE

Before looking at the causes of incontinence, it's important to understand how urine is normally stored and released from the body (see Figure 8.1). First, the kidneys filter out wastes from the blood to form urine. That urine

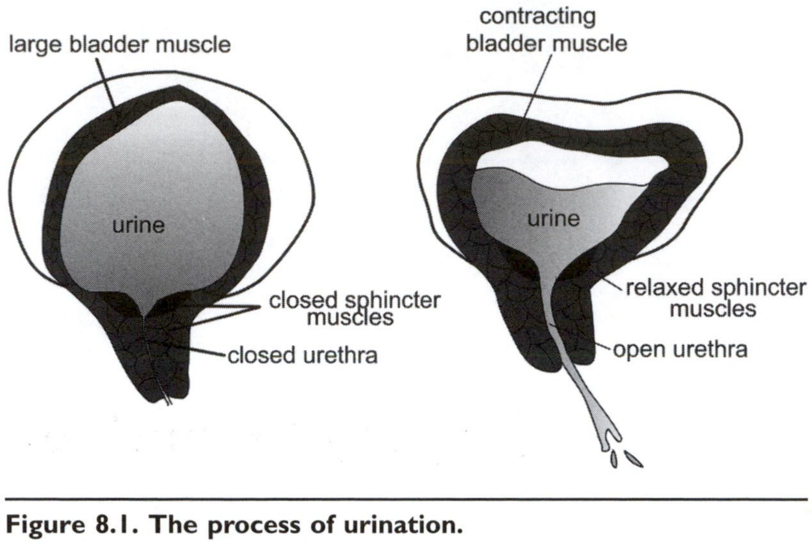

Figure 8.1. The process of urination.
As the bladder fills, the muscles in its walls relax and the sphincter muscles remain closed to prevent leakage. When the bladder is ready to empty, its muscles contract, and the sphincter muscles relax to release the urine.

flows down through the ureters and into the bladder, where it is stored until an appropriate time for evacuation. The bladder is a balloon-like organ, which expands to accommodate the growing volume of liquid. As the bladder fills, the detrusor muscles in its walls remain relaxed and the internal sphincter muscle, which encircles the top of the urethra at the bladder neck, contracts to prevent urine from leaking.

When the amount of urine in the bladder rises to the point of fullness, receptors in its walls send a message to the spinal cord via sensory neurons. The spinal cord passes the message along to the brain in the form of impulses, which causes the feeling of pressure in the lower abdomen that signals the need to urinate. At the appropriate time (i.e., when the person has reached a bathroom), parasympathetic nerves release the neurotransmitter acetylcholine, which causes the detrusor muscle in the wall of the bladder to contract and the internal urethral sphincter to relax. The bladder contraction forces urine through the urethra. But before the stream can exit the body, the individual must consciously relax the external urinary sphincter and pelvic floor muscles.

Any damage or disruption to the muscles, nerves, or psychological functions that control urination can lead to incontinence (see Causes of Incontinence in Table 8.1). For example, injuries to the spinal cord or neurological diseases such as multiple sclerosis, muscular dystrophy, or stroke can interrupt the signals to and from the bladder and the brain, causing the bladder to contract at inappropriate times. Weakening of the bladder muscles,

TABLE 8.1. Causes of Incontinence

Medical Conditions	Bladder cancer Birth defects Hole in bladder (fistula) Urinary stones Interstitial cystitis Bladder outlet obstruction (BPH, prostate tumor) Diabetes Nerve disorders (multiple sclerosis, stroke) Psychosis (dementia)
Medications	Narcotic analgesics Sedatives Anticholinergics (antihistamines) Antipsychotics Antidepressants Alcohol
Aging	Decline in bladder capacity Decrease in urethral closure pressure (in women) Loss of sphincter strength Loss of bladder elasticity Detrusor muscle overactivity
Surgery	Cesarean section Prostatectomy

the pelvic floor muscles that support the bladder, or the sphincter muscles can also result in unplanned urine loss. Dementia and other forms of psychosis can prevent a person from interpreting the signal that it is time to urinate.

Although incontinence is not a natural result of the aging process, the elderly are more susceptible because of changes to their urinary tract. With age, the bladder begins to lose its elasticity as well as its ability to hold large quantities of liquid. The detrusor muscles often overact, meaning that they contract before the bladder is full. In women, the muscles that hold the bladder in place as well as the sphincter muscles degenerate as a result of stretching during childbirth and/or the loss of estrogen (which keeps tissues firm) following menopause. As men age, their prostate gland begins to grow (called benign prostatic hyperplasia) and often blocks the urethra, through which urine must pass to exit their bodies. The blockage forces urine to back up and eventually overflow, to the point where it continually leaks from the bladder.

Medications such as antidepressants, antihistamines, sedatives, and antipsychotic drugs may either cause or exacerbate incontinence by interfering with the nerve signals and muscle contractions necessary to the urination process. Diuretics (including the caffeine in soft drinks and coffee) increase urine production by the kidneys, causing the bladder to fill more quickly than usual. Alcohol dulls the nerves in the bladder, making it difficult for the brain to sense fullness.

TYPES OF INCONTINENCE

There are several different types of incontinence: urge, stress, mixed stress and urge, overflow, functional, and structural (see Table 8.2).

Urge Incontinence

Popularly known as overactive bladder, urge incontinence is the sudden need to urinate, often followed by an involuntary release of urine. The patient feels that he or she desperately needs to reach a toilet, but is unable to do so in time. Whereas a healthy individual will urinate anywhere from four to six times a day, a person who has an overactive bladder may need to go more than eight times a day (called frequency) and may need to wake at night to urinate or fail to wake and release urine in his or her sleep.

TABLE 8.2. Types of Incontinence

Type of Incontinence	Description	Treatment
Urge	Sudden need to urinate, followed by involuntary loss of urine	Antibiotics (if caused by infection), bladder retraining, pelvic floor muscle exercises, diet, medication, electrical stimulation of the sphincter or sacral nerve, surgery
Stress	Involuntary leakage of urine when pressure is placed on the abdomen	Pessary to support prolapsed bladder, medication, surgery
Mixed stress and urge	An overactive bladder muscle, coupled with weakened pelvic floor and sphincter muscles	One or more of the treatments indicated for urge and stress incontinence
Overflow	A blockage in urine flow, or weakened bladder contractions that prevent the bladder from emptying completely	Surgical removal of the blockage, short-term catheterization
Functional	A physical or mental disability that prohibits the patient from reaching a bathroom in time to urinate	Treating the cause of the disability
Structural	A hole or other defect that causes urine to leak out of the bladder	Surgical correction of the defect

Urge incontinence usually occurs when nerve damage interferes with messages traveling from the bladder to the brain and vice versa. The detrusor muscle in the bladder contracts at inappropriate times, expelling urine before a person is ready to urinate. Neurological disorders such as stroke, multiple sclerosis, Alzheimer's, and Parkinson's can interfere with these nerve pathways. Lower urinary tract infections may also cause urge incontinence because they irritate the bladder lining and reduce the volume that the bladder can comfortably hold.

Stress Incontinence

Stress incontinence is the involuntary leakage of urine any time pressure is placed on the abdomen, for example by coughing, sneezing, laughing, or lifting a heavy object. Women are more likely to suffer from stress incontinence, because stretching of the muscles during childbirth and estrogen deficiency following menopause weaken the sphincter muscles as well as the pelvic floor muscles that support the bladder. Weakness in the pelvic floor muscles can allow the bladder to droop downward toward the vagina (called prolapse or cystocele). When the bladder drops, the sphincter muscle is unable to completely seal off the opening to the urethra, especially if it too has been weakened. So any time pressure is placed on the abdomen, the sphincter opens like a leaky valve and releases a few drops or a full stream of urine.

Mixed Stress and Urge Incontinence

Nearly a third of women with incontinence suffer from a combination of stress and urge incontinence. This condition combines an overactive bladder muscle with weakened pelvic floor and sphincter muscles that are unable to stop the sudden release of urine.

Overflow Incontinence

If a blockage inhibits the passage of urine from the bladder through the urethra, or if bladder contractions are not powerful enough to force urine out of the body, the result is overflow incontinence. Instead of releasing in a powerful stream, urine dribbles out and the bladder is never able to fully empty. The bladder eventually becomes so full that it overflows and constantly leaks urine. Unlike stress incontinence, which primarily affects women, overflow incontinence is more common in men. As the prostate enlarges with age, it presses against the urethra, blocking the passageway by which urine would normally leave the body. Tumors and urinary stones may also block the urethra, causing similar results. Weakened bladder contractions may come as a result of nerve damage from diabetes, excessive alcohol consumption, or neurologic diseases.

Functional Incontinence

A patient who suffers from functional incontinence may have a perfectly healthy urinary system, but because of disease or physical impairment is unable to reach the bathroom in time to urinate. Functional incontinence is common among the elderly, especially when mental capacity is compromised by Alzheimer's or dementia, or motility is restricted by stroke, weakness, arthritis, or poor vision.

Structural Incontinence

A hole (called a fistula) between the bladder and the vagina (vesicovaginal fistula) or between the urethra and the vagina (urethrovaginal fistula) can cause constant leakage of urine (see Chapter 6 for more on fistulas). A fistula may result from difficult childbirth, pelvic surgery (hysterectomy or bowel resection for diverticulitis), or radiation therapy for cancer. Birth defects and other structural abnormalities of the bladder and urethra may also result in urine leakage.

EVALUATION FOR INCONTINENCE

Evaluation begins with a visit to a urologist or urogynecologist (for women), both of whom specialize in urinary problems and conditions. To diagnose the root of a patient's incontinence and determine what type of incontinence is causing the problem, the doctor must take a detailed medical history, followed by a physical exam, laboratory, and other diagnostic tests.

Medical History

During an office visit, the doctor will assess a number of factors:

- Length and severity of symptoms
- When leakage occurs (for example, only during lifting, sneezing, or coughing, or throughout the day)
- Family history of incontinence
- Medications the patient is taking
- Patient history of neurologic disease (for example, multiple sclerosis)

Voiding Diary

Often the patient can provide enough details to confirm a diagnosis, but sometimes the doctor will request more information in the form of a voiding (urination) diary (see Table 8.3). For a period of twenty-four hours to a week, the patient will record his or her fluid intake, the time and volume of each urination, symptoms (such as a strong urge to urinate, and pain or

TABLE 8.3. Sample Voiding Diary

Date	Time	Drinks	Leakage Volume	Urge to Go? (Yes/No)	Activity
Feb. 3	9:00 A.M.	1 cup orange juice	2 ounces	No	Sneezing
Feb. 3	3:30 P.M.	2 cans soda	5 ounces	Yes	None
Feb. 5	2:00 A.M.	1 cup milk before bed	3 ounces	No	Sleeping
Feb. 5	10:00 A.M.	2 cups coffee	8 ounces	Yes	Jogging
Feb. 6	8:00 A.M.	1 cup orange juice	2 ounces	Yes	Laughing
Feb. 6	2:00 P.M.	2 cups apple juice 1 cup coffee	7 ounces	No	Sneezing

pressure associated with urination or the urge to urinate), and leakage episodes.

Physical Examination

Next, the doctor will examine the patient, searching for conditions that could contribute to or cause incontinence. For example, the discovery of poor reflexes could indicate an undiagnosed neurologic disease. Lumps or hard masses in the abdomen could be signs of a tumor or enlarged bladder. The doctor will also check the patient's sphincter control via a digital rectal exam, asking the patient to squeeze and release the sphincter muscle as if starting and stopping the flow of urine. From here, the examination will differ for men and women. In men, the doctor will also use the digital rectal exam to check for prostate enlargement and tumors. In women, the doctor will examine the vagina and pelvis, looking for a fallen bladder or weakness of the pelvic floor muscle.

Urine and Blood Tests

Because a urinary tract infection or tumor can cause symptoms of urgency and frequency, the doctor will check the patient's urine under a microscope for the presence of white or red blood cells, which would, respectively, indicate one of these two conditions. If the doctor suspects diabetes or a similar condition, he or she may also check the blood for urea nitrogen (BUN) and creatinine levels to see whether the kidneys are adequately filtering wastes.

Cystoscopy

A procedure known as cystoscopy (see Chapter 6) can provide the doctor with valuable clues on the source of a patient's incontinence. By inserting a thin flexible tube called a cystoscope through the urethra and into the bladder, the doctor can look for abnormalities such as stones, tumors, or bladder lining irritation.

Postvoid Residual Measurement

To determine whether urine remains in the bladder after the patient urinates, the doctor measures the postvoid residual volume. After the patient has urinated, the doctor inserts a catheter into the bladder, drains, and measures the amount of urine remaining. A volume of less than 50 milliliters is considered normal, whereas more than 200 milliliters can signal a bladder-emptying problem.

Stress Test

Primarily designed to diagnose stress incontinence in women, this test can indicate whether pressure on the abdomen is causing urine to leak out through the urethra. The test is simple: The patient coughs vigorously, while the doctor examines the vaginal area for signs of urine loss.

Urodynamic Evaluation

A **urodynamic evaluation** is a series of tests that determine the type of voiding dysfunction by gauging how well a patient's bladder, urethra, and rectum are holding and releasing urine. Using a variety of techniques, the doctor measures bladder pressure and analyzes urine flow during bladder filling and emptying. Tests include the following:

Cystometry. This test measures the pressure inside a patient's bladder as well as the volume of liquid the bladder can hold. As the bladder is filled with liquid via a catheter, the patient is asked to relate sensations of coolness, wetness, or fullness. A device called a cystometer electronically measures bladder pressure during filling. Once the patient feels that his or her bladder cannot hold any more liquid, the doctor records the volume of liquid that has been inserted as the maximum bladder capacity.

Uroflometry. The patient stands or sits and urinates into a funnel. While the patient is urinating, a machine electronically calculates the flow rate and the amount of urine voided. If the flow is weak, more tests are needed to determine whether the problem stems from an obstruction or weak bladder contractions.

Sphincter Electromyography. The doctor places a patch on either side of the patient's rectum or inserts a needle electrode just above the rectum in men and next to the urethra in women. The machine measures the strength of each sphincter muscle contraction during bladder filling and emptying on a machine called an electromyograph (EMG).

Synchronous Multichannel Videourodynamic Studies. This high-tech device measures bladder and rectum pressure as well as urinary flow rate. The recorded information is displayed as a series of graphs alongside an x-ray of the bladder and urethra. Unfortunately, this effective diagnostic tool is not yet available in every hospital.

TREATMENT OPTIONS

Once incontinence was viewed as an inevitable part of the aging process; a permanent annoyance that patients just had to learn to live with. Now, as doctors gain a greater understanding of the underlying problems that cause incontinence and as patients have broader access to informational materials relating to incontinence, the condition is becoming increasingly easier to treat.

A majority of patients can be treated with lifestyle changes alone. Sometimes, simple dietary alterations or the elimination of certain medications can easily solve the problem. More often, patients need to retrain their bladder or strengthen the muscles that support the bladder. If behavioral modifications alone aren't effective, doctors may try medication, surgery, or a combination of all three. Regardless of the cause, doctors will usually begin with behavioral techniques, because they are the least invasive and tend to yield the greatest results.

Treatments for Urge Incontinence

If a patient's incontinence is caused by an underlying infection, doctors will treat it with antibiotics, which usually clear up the problem. Likewise, if bladder or prostate cancer is to blame, treatment is necessary to resolve these diseases and often to save the patient's life (see Chapters 6 and 7 for more on bladder and prostate cancer treatments).

An overactive bladder is not curable, but it is manageable through behavioral techniques, medications, or, as a last resort, surgery.

BLADDER RETRAINING

This technique is a mainstay for most types of incontinence, as it helps patients regain control over their bladder and sphincter muscles. To begin, the patient schedules specific times for voiding, for example once every two hours. Gradually, the patient increases the time between each bathroom visit, using distraction and relaxation techniques to suppress the urge to urinate. If the technique is effective, patients can increase their bladder capacity while gaining control over their sphincter muscles.

PELVIC FLOOR MUSCLE EXERCISES

These exercises, called Kegels after the gynecologist who named them, strengthen the pelvic floor and sphincter muscles. They are often used by women who either are pregnant or have recently delivered to maintain muscle tone through the natural stretching that occurs during pregnancy. The pelvic floor muscles act like a hammock to support the bladder, and the sphincter prevents the urethra from opening and expelling urine. Strengthening these muscles not only keeps the urethra closed but also prevents involuntary bladder contractions. When the pelvic floor muscles contract, the detrusor muscles of the bladder naturally relax, preventing urine leakage through the urethra. To do a Kegel, the patient squeezes and releases as if starting and stopping the flow of urine. Often, biofeedback is used at the start of therapy to help the patient isolate the correct muscles. A small device is inserted in the vagina or rectum, which displays the contractions on a monitor, allowing the patient to see the strength of each muscle contraction. The patient holds each contraction for 5–10 seconds and repeats each set of ten several times, building up to about 30–60 Kegels per day. Patients usually see improvement after six to eight weeks.

DIET

One of the simplest modifications to implement is dietary. If a patient cannot hold his or her bladder overnight, a doctor may recommend that the patient limit or avoid fluids before bedtime. Patients will also want to limit diuretics such as caffeine, which increase urine production, as well as spicy foods, chocolate, and nicotine, all of which can irritate the bladder.

MEDICATION

If behavioral modifications are ineffective, the doctor may prescribe one of several medications designed to treat urge incontinence. The most commonly prescribed are anticholinergics such as oxybutynin (Ditropan) and tolterodine (Detrol). These medications block the action of the neurotransmiter acetylcholine, which stimulates bladder contractions. Patients taking these two medications have reported side effects such as dry mouth, constipation, blurred vision, and drowsiness, but newer time-release versions have fewer side effects and allow for more convenient once-a-day dosing. Tricyclic antidepressants such as imipramine (Tofranil) also inhibit detrusor muscle contractions and may be used alone or in conjunction with anticholinergics.

Older women with urge or stress incontinence may benefit from estrogen therapy, which corrects muscle atrophy resulting from postmenopausal estrogen loss. The long-term benefits are still unproven, however, and because of the increased uterine and breast cancer risk, many doctors are reluctant to prescribe estrogen therapy for their patients.

ELECTRICAL STIMULATION OF THE SPHINCTER OR PELVIC FLOOR MUSCLE

Doctors place an electrode in or near the patient's vagina or rectum that stimulates the pelvic floor or sphincter muscle to contract (much like a Kegel exercise).

SACRAL NERVE ROOT STIMULATOR

A tiny wire implanted near the base of the spine is attached to a pacemaker-like device that delivers mild impulses to the patient's sacral nerve. This nerve controls the bladder, rectum, and pelvic floor muscles. The electric stimulation "shuts off" symptoms of urgency and frequency, and eases pelvic pain.

SURGERY

As a last resort, when a patient has not responded to behavioral modifications or medication, doctors try to correct urge incontinence surgically. The most common and most successful method used is augmentation cystoplasty. Surgeons use part of the patient's intestine to augment the bladder, increasing its capacity and preventing involuntary contractions. The success rate is high—about 90 percent—but many patients report having difficulty urinating on their own after the procedure.

Until an effective treatment is found, patients may decide to use absorbent pads or undergarments to prevent embarrassing leaks and hold wetness away from their body.

Treatments for Stress Incontinence

The first line of treatments for stress incontinence follow the same philosophy as those designed for urge incontinence. The goal is the same for both conditions: Strengthen the pelvic floor muscles that support the bladder as well as the sphincter muscles that prevent urine leakage. As with urge incontinence, behavioral treatments such as pelvic floor exercises, biofeedback, and dietary adjustments are always attempted first.

If stress incontinence is caused by a cystocele (bladder prolapse), which is often the case in older women, doctors insert a plastic device called a pessary (see Figure 8.2) to support and raise the bladder neck. The pessary presses against the wall of the vagina and urethra, repositioning the urethra to prevent leakage. This device is a less invasive treatment than surgery, but because of the infection risk of having a foreign object in the vagina, women must be conscientious about removing and cleaning their pessary at least two to three times per week.

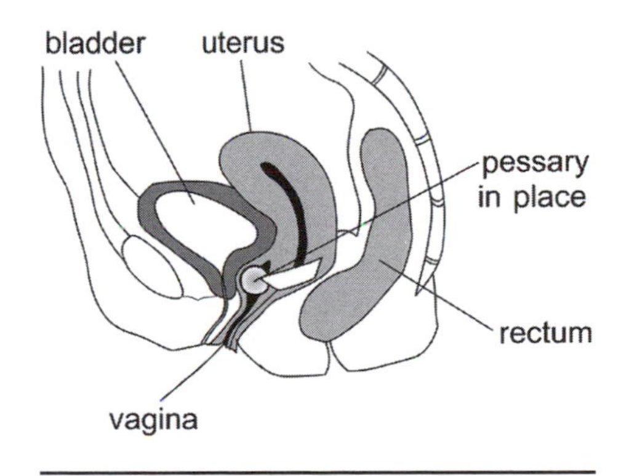

Figure 8.2. A pessary.
The pessary is a plastic device that supports and raises the bladder neck when stress incontinence is caused by bladder prolapse.

MEDICATION

When behavioral modifications and simple devices such as the pessary are not enough to prevent leakage, doctors may prescribe medications to hold the sphincter closed. The body normally produces neurotransmitters, or chemical messengers, that force the sphincter to remain closed. Drugs called alpha-adrenergic agonists (ephedrine, pseudoephedrine, and phenylpropanolamine [PPA]), found in over-the-counter cold medications and appetite suppressants, recreate the effects of neurotransmitters and increase sphincter muscle tone. But these medications come with serious side effects. They may cause high blood pressure, irregular or rapid heartbeat, anxiety, and insomnia. In 2000, following a warning by the U.S. Food and Drug Administration (FDA) that PPA was associated with strokes, manufacturers of cold medications began removing the drug from their products.

An alternative to alpha-adrenergic agonists is tricyclic antidepressants, which are also used to treat urge incontinence. The antidepressants amitriptyline (Elavil) and imipramine (Tofranil) relax the bladder and strengthen the internal sphincter muscle. Estrogen therapy has also been used to correct vaginal atrophy in women suffering from both stress and urge incontinence, but there is no firm evidence that it is beneficial over the long term.

SURGERY

For the most severe cases of stress incontinence, where all other treatments have failed, surgery is the most effective way to restore bladder and urethral support and prevent leakage. The most popular procedures either raise the bladder or support it with a "sling" (see Figure 8.3) made from the patient's own tissues. In retropubic suspension procedures, the surgeon lifts the patient's bladder and secures it with string to the pelvic bone or a nearby muscle. These operations require a wide incision in the abdomen or vagina, but newer, laparoscopic techniques allow for smaller incisions and thus less invasive surgery.

In sling operations, doctors use a band of the patient's tissue (usually a

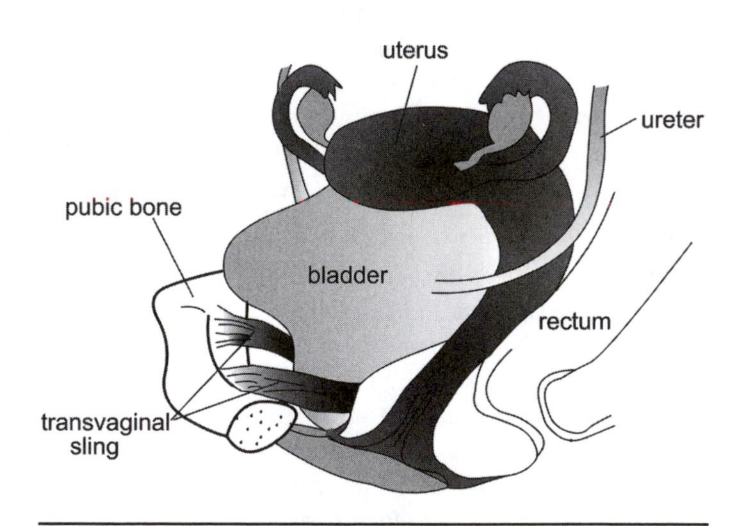

Figure 8.3. Bladder sling procedure.
Doctors create a sling out of a band of the patient's tissue, which supports the bladder neck.

layer of muscle tissue called the fascia) to fashion a hammock on which the bladder neck is supported. In a newer, less invasive form of the sling procedure called tension-free vaginal tape (TVT), the surgeon attaches polypropylene mesh tape in a U shape around the urethra. The tape forms a sling, which is tightened to the point where it prevents leakage but still allows the patient to urinate normally.

In patients whose sphincter muscle is damaged, doctors may implant a doughnut-shaped artificial sphincter around the bladder neck to hold it closed. The artificial sphincter remains closed until the patient wants to urinate, at which time he or she compresses a valve implanted under the skin that releases urine from the bladder. There are two drawbacks to this procedure: The artificial sphincter may malfunction, requiring a second operation; or the device may become infected and need to be removed.

A less invasive procedure for patients with sphincter dysfunction involves the injection of a bulking agent, such as human or animal collagen, into the tissues surrounding the urethra. The collagen builds up the urethral tissues, reducing leakage and helping to hold the urethra closed. The downside to this procedure is its short lifespan. Over time, the patient's body begins to absorb the injected materials, and the surgery must be repeated.

Treatments for Overflow Incontinence

Doctors usually treat overflow incontinence by surgically removing the source of the blockage. In men, doctors remove the enlarged part of the prostate via one of two procedures: transurethral resection of the prostate (TURP), removing the obstructing tissue one piece at a time; and transurethral incision of the prostate (TUIP), widening the urethra through a few small incisions (for more on these procedures, see Chapter 7). Men with BPH may also benefit from drugs called alpha-adrenergic blockers (doxazosin, prazosin, terazosin, or tamsulosin), which improve urine flow. In women whose overflow incontinence is caused by a fallen bladder, surgery to raise and support the bladder is usually most effective.

If a patient's overflow incontinence results from detrusor muscle dysfunction, short-term bladder catheterization may be necessary to remove urine until the muscle begins functioning properly. The patient inserts a soft tube through the urethra into the bladder (via the penis in men and the vagina in women) to drain the urine into a bag attached to his or her leg. The catheter may be permanently implanted in the patient's body or inserted only as needed.

Treatments for Functional Incontinence

Doctors begin by identifying and treating the underlying cause of functional incontinence, for example by prescribing antipsychotic drugs for dementia or antibiotics for a urinary tract infection. Until the treatment has taken effect, patients may benefit from scheduled voiding, catheterization, or absorbent pads.

BEDWETTING (NOCTURNAL ENURESIS)

When babies are born, they have no conscious control over the urination process. When the bladder is full, it automatically contracts and the sphincter opens, releasing a flow of urine into the baby's diaper. This unconscious voiding is called the micturition reflex. As babies grow, their nervous systems mature, and they become aware of signals indicating that their bladder is full. Eventually, babies are able to distinguish the need to urinate and respond to that need by using the toilet or contracting the sphincter muscle to delay urination until an appropriate time. The physical maturation needed to receive messages and urinate in the toilet is usually reached at around age 2 (2½ for boys), but it may take months or even years longer for some children to stay dry through the night.

About 10 percent of children over the age of 5 continue to wet the bed, according to the National Association for Continence (2002). Most bedwetters grow out of their problem by the age of six or so, but some continue into adolescence. There are two types of nocturnal enuresis: primary enuresis, in which a child never successfully toilet trains and continually wets the bed; and secondary enuresis, in which the child initially had dry nights after toilet training, but began wetting the bed because of an underlying physical or psychological problem. Wetting most often occurs when the child is in rapid eye movement (REM) sleep—the deepest sleep cycle—when brain control over urination is lost.

What Causes Bedwetting?

The roots of bedwetting may be physical, for example a bladder that cannot hold enough liquid, an overproduction of urine, or an overactive bladder. Or it may be psychological, caused by the stress of a new move or the birth of a sibling. Some children are born with a urinary defect that inhibits normal urination, such as a bladder or urethral blockage or a neurologic disease like spina bifida. Bedwetting may also have a genetic component. If one parent wet the bed, his or her child runs about a 50 percent chance of also being a bedwetter. If both parents were bedwetters, the odds jump to 80 percent (National Kidney and Urologic Diseases Information Clearinghouse, "Urinary Incontinence in Children," 2001).

A final theory is that children make more urine at night because of an abnormality in the secretion of antidiuretic hormone (ADH), which regulates urine production. The pituitary gland produces ADH in a circadian rhythm, reaching peak production at around the same time every night. An increase of ADH signals the kidneys to cut back on urine production. Studies have indicated that many children have a reversal in this natural rhythm, producing less ADH at night, which in turn stimulates the kidneys to produce more urine than the bladder can hold through the night.

Diagnosis and Treatment of Bedwetting

When a child arrives at his or her pediatrician's office with a bedwetting problem, the first thing the doctor will do is attempt to rule out an underlying medical condition, for example a urinary tract infection (the most common cause of secondary enuresis), congenital abnormalities such as vesicoureteral reflux, or a neurologic disease. If a urinalysis reveals a urinary tract infection, the doctor will prescribe antibiotics, which should clear up the problem within a few days or weeks. Congenital abnormalities are more tricky to treat and often require invasive surgery.

Most bedwetting eventually resolves itself as the bladder grows and the child becomes more aware of the signals governing the urination process. In the meantime, simple behavioral modifications can prevent bedwetting episodes. First, parents can limit the amount of fluids their child drinks before bedtime, especially diuretics like sodas, chocolate, and tea. "Double-voiding," or going to the toilet 20 to 30 minutes before bedtime and then again at bedtime, fully empties the bladder before the child goes to sleep. To help a child expand his or her bladder capacity, a doctor may recommend that the child keep a daily voiding diary. In the diary, the child writes down the times at which he or she normally uses the toilet. Using the diary as a guide, the child tries to gradually increase the time between each urination by about 15 minutes per week. If none of these methods is successful, parents often turn to alarm systems, which wake up the child when he or she starts to urinate. These devices, which contain a water-sensitive pad worn inside the child's pajamas, may not help a child get to the bathroom in time to prevent an accident, but they can increase the child's awareness of when he or she is wetting the bed.

When behavioral modification alone is not enough, doctors may prescribe one of several medications to treat bedwetting. If the cause of the problem is an overactive bladder, anticholinergics such as oxybutynin (Ditropan), imipramine (Tofranil), or hyoscamine (Levsin) relax bladder muscle contractions. These medications are successful in up to 80 percent of children. Side effects include dry mouth and increased sun sensitivity in fair-skinned children. Medications such as desmopressin (DDAVP) increase ADH levels to suppress urine production at night. DDAVP is administered in a nasal spray just before bedtime. Adverse reactions may include stuffy nose, headaches, and abdominal cramps.

NOCTURIA (NOCTURNAL ENURESIS) IN ADULTS

Whereas urinating during the night is normal for young children, in adults, getting up more than once or twice a night or wetting the bed is usually the sign of a more serious health problem. It may be that the kidneys are producing more urine than the bladder can hold because the person has ingested too many fluids before bedtime or has taken a diuretic. Or, bladder capacity may be reduced from an enlarged prostate, bladder or prostate cancer, or interstitial cystitis. Another possibility is that a neurologic condition, such as Parkinson's disease or multiple sclerosis, is causing involuntary bladder contractions.

To diagnose the problem, a urologist will first complete a medical history, noting duration of symptoms, presence of pain during urination or a weak urine stream, and tingling or weakness in the arms and legs (which could indicate a neurologic condition). Next, the doctor will check the patient's urine for blood, a sign of a urinary tract infection. The doctor may also ask the patient to fill out a voiding diary for one to two days, and compare daytime versus nighttime urine production to determine whether the problem is excessive urine output or small bladder capacity. The final stage of the examination will most likely include uroflometry, postvoid residual measurement, and urodynamic evaluation to measure the urine flow rate, volume of urine voided, volume remaining in the bladder after urination, and bladder and sphincter pressure.

Depending on the underlying cause of the dysfunction, nocturia is treatable with medications, behavioral modification, catheterization, or surgery. Doctors treat infections with antibiotics. Prostate blockages in men are treated with alpha-adrenergic blocking agents like terazocin (Hytrin) or doxazocin (Cardura), which improve urine flow. In cases where medication does not clear the blockage, TURP or TUIP surgery can reduce the size of the prostate and open up the urethra. If the problem is small bladder capacity, the medication DDAVP can reduce the amount of urine produced by the kidneys. Finally, anticholinergics can prevent involuntary bladder muscle contractions. Patients may also want to drink fewer fluids before bedtime, especially diuretics such as sodas, coffee, and tea.

URINARY RETENTION

In many cases, the involuntary release of urine is not the problem, but rather the inability to produce an adequate urine flow or any urine at all. Urinary retention may take many forms, ranging from a weak urine stream to the inability to urinate.

Hesitancy. A patient who goes to the bathroom and must wait for a few seconds before initiation of a urine stream suffers from hesitancy. Hesitancy may not always indicate a problem, especially if the person is attempting to urinate before his or her bladder is full.

Thin or Weak Stream. A blockage in the urethra (usually caused by an enlarged prostate in men), weak bladder contractions, or an inability to properly relax the sphincter muscles can result in a weak urine stream.

Intermittency. Instead of flowing out in a steady stream, urine is released in spurts. Intermittency may be caused by a conscious or unconscious tightening and releasing of the sphincter muscle during urination, or from a neurologic condition called detrusor-sphincter dyssynergia, which usually follows a spinal cord injury or multiple sclerosis.

Postvoid Dribbling. When a person stands up after urinating, leftover urine dribbles out of the penis or vagina. In men, urine may remain in the urethra after voiding because of a prostatic obstruction. In women, urine can remain trapped inside the vagina because of insufficient wiping.

Inability to Urinate. It can be a frightening experience to suddenly lose the ability to urinate. Despite a painful urge, patients are unable to go once they reach the bathroom. Men may also feel a burning sensation or pressure at the tip of their penis. The problem may be caused by a blockage in the urethra (most likely an enlarged prostate in men), a failure of the bladder to contract because of a spinal cord injury or neurologic condition, or a stretching and weakening of the bladder muscle following childbirth or surgery.

Urinary retention can come on suddenly, or it may develop over a period of months or even years. It usually results from one of three things: a blockage in the urethra, a weakened bladder that is unable to contract strongly enough to empty itself, or damage to the nervous system that prevents bladder contractions.

Doctors analyze the problem by first conducting a medical history and physical exam, and by checking the urine for signs of an infection. If the urinalysis is negative, the next step is uroflometry to measure the strength of urine flow, and a postvoid residual urine volume to determine how well the bladder is emptying itself.

Treatment depends on the type of problem that is causing urinary retention. Urinary tract infections are treated with antibiotics, and blockages in the urethra are treated with medication or surgery, as are neurologic conditions. Self-intermittent catheterization (SIC), or draining urine into a bag via a flexible tube inserted in the bladder, is a temporary way to manage the problem until other treatments take effect.

Urinary Tract Infections

The human body is a well-armed fortress, protected by natural barriers and an army of white blood cells poised and ready to combat invading troops of bacteria and viruses. But on occasion, the enemy is able to slip by the body's defenses and launch a full-scale attack. This attack can occur virtually everywhere in the body; but the urinary system, because of its easy access route to the outside, is especially vulnerable.

To protect itself against attack, the urinary system has several natural safeguards in place. Although urine contains waste products, it is sterile and free from bacteria, viruses, and other microorganisms. Any bacteria that do slip in are flushed out of the body when the bladder empties fully during urination. The path of urine flow is specially designed with valves that prevent urine from backing up from the bladder into the kidneys or from the urethra into the bladder, where it might stagnate and collect bacteria. In men, the urethra is long enough (around 8 inches) to make it difficult for bacteria to reach the bladder. The man's prostate gland also produces secretions that slow bacterial growth. In women, who have shorter urethras (around 1.5 inches), the vaginal opening is guarded by beneficial microorganisms called lactobacilli, which inhibit the colonization of more harmful bacteria.

But despite these safeguards, the entryways to the urinary system (the vagina, penis, and rectum) often harbor bacteria, which can sometimes make their way into the urinary system and cause an infection. Once an infection has reached the urethra, it can spread to the bladder or kidneys. In rare cases, bacteria can also enter the urinary tract through the bloodstream, carried from an infection elsewhere in the body.

According to the American Foundation for Urologic Disease (2001), uri-

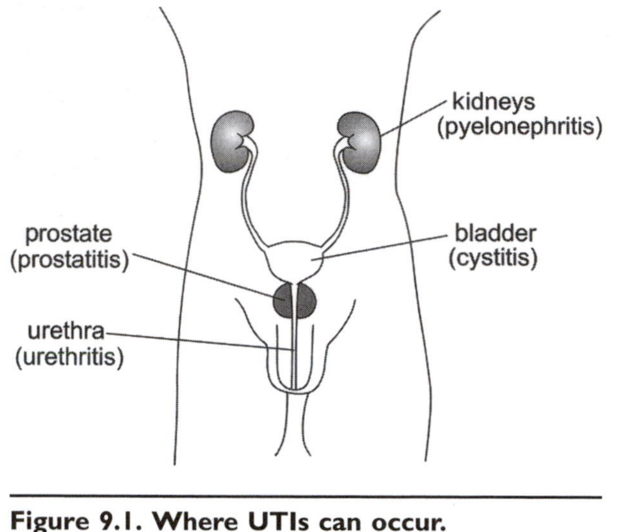

Figure 9.1. Where UTIs can occur.
Bacteria can cause infections in the kidneys, bladder, urethra, and prostate (in men).

nary tract infections (UTIs) account for between 8 and 10 million doctor's visits every year. Women are far more likely to develop an infection than men, in part because their short urethra provides easier access to their bladder and also because their urethral opening is close to their anus and vagina, where bacteria are known to breed. For these reasons, one in five women can expect to develop a UTI at some point in their lives. Older men are also susceptible, because the prostate enlargement that occurs as men age (see Chapter 7) can block urine flow. Urine left stagnating is likely to breed infection-causing bacteria.

Infections are named based on their location within the urinary system: the bladder (cystitis), the urethra (urethritis), the prostate (prostatitis), or the kidneys (pyelonephritis) (see Figure 9.1). Cystitis is the most common UTI, especially among women. Pyelonephritis can be the most dangerous, resulting in permanent kidney damage if left untreated. More than 90 percent of infections are caused by *Escherichia coli* (*E. coli*) (see photo in color insert), a bacteria that normally resides in the intestinal tract but can easily spread from the opening of the anus into the urethra. In some cases, sexually transmitted bacteria such as chlamydia and mycoplasma can also cause UTIs, but these types of infections are usually confined to the urethra and reproductive system.

Risk Factors for Urinary Tract Infections

Women are more likely than men to develop a UTI, because their short urethra provides easy access for bacteria to reach the bladder. Some other common risk factors include:

Blockage. A stone, enlarged prostate, or other blockage in the urinary tract forces urine to pool and stagnate, where it becomes a breeding ground for bacteria.

Disease. Diabetes and other diseases that suppress the immune system can inhibit the body's ability to fight off an infection.

Congenital Abnormalities. Babies who are born with abnormalities of the urinary tract that inhibit the flow of urine or force urine to back up into the bladder or kidneys are at increased risk for UTIs.

Intercourse. Many doctors believe that friction during sexual intercourse can force bacteria from the man's penis into the woman's urethra. Women who are sexually active are more susceptible to sexually transmitted diseases like chlamydia, gonorrhea, or herpes simplex virus, all of which have been linked to UTIs. Studies have also found that women who use a diaphragm with spermicide or whose partners use condoms containing a spermicidal lubricant are more likely to develop an infection, although the reasons for the link are unclear.

Catheterization. Catheters, which are used for urine removal in patients who are incontinent (see Chapter 8), can breed bacteria if not removed and cleaned regularly. The longer the catheter is in place, the greater the risk of infection.

Pregnancy. Experts say that as many as 2 to 4 percent of pregnant women are likely to develop a UTI. A shift in bladder position during pregnancy provides an easy route through which bacteria can travel up the ureters to the kidneys. UTIs are particularly dangerous during pregnancy, especially if the infection reaches the kidneys, because they can trigger premature labor. Because of the risks involved, most obstetrician/gynecologists take routine urine samples throughout their patients' pregnancies.

Menopause. When a woman goes through the "change," her body produces less estrogen, which is crucial for vaginal wall strength, and for maintaining the protective mucous membrane that prevents bacteria from entering the vagina.

Symptoms of a Urinary Tract Infection

In some cases UTIs will arrive without any physical indication, but generally infections reveal themselves through one or more of the following symptoms:

- A frequent urge to urinate, but very little urine is released
- Pain or burning in the bladder or urethra when urinating (pain during urination is called dysuria)
- A feeling of pressure or fullness above the pubic bone in women, and around the rectum in men
- Aching in the prostate or testicles or pain during ejaculation if the prostate is infected (in men)
- Cloudy, bloody, or foul-smelling urine
- Flu-like symptoms; for example, fever, chills, nausea, or vomiting if the infection has spread to the kidneys
- Pain in the back or side below the ribs, which is also a sign of a kidney infection
- Pus in the urine or a greenish-yellow or white discharge from the penis or vagina

Diagnosing a Urinary Tract Infection

Because the symptoms of a UTI can mimic those of interstitial cystitis (a bladder irritation—see Chapter 6), kidney or other urinary tract stones, or cancerous tumors, doctors examine patients' urine for evidence of bacteria, which would rule out these other conditions. So that the urine sample is not contaminated by bacteria living on the outer genitals, the patient is typically asked to provide a "clean-catch" sample. This involves first washing the genital area with soap and water, then collecting the urine in a sterile container in midstream. The doctor examines the urine under a microscope for the presence of red and white blood cells and bacteria, any of which would indicate an infection. If the urine sample reveals an infection, the doctor will also do a urine culture, growing the bacteria in a laboratory for a 24–48-hour period. During that time, the bacteria will tested against various antibiotics (called a sensitivity test) to determine which one will work most effectively.

When a patient has recurrent UTIs or when a single UTI does not clear up with antibiotics, the doctor will probably call for one or more of the following imaging tests to rule out urinary system abnormalities, stones, and tumors:

Intravenous Pyelogram (IVP). A special contrast dye injected into a vein outlines the structure of the bladder, kidneys, and ureters on x-ray film. The x-ray reveals any structural abnormalities or blockages that are preventing the kidneys from properly draining.

Ultrasound. The same sound wave technology used to view a growing fetus in its mother's uterus allows doctors to visualize the kidneys and bladder.

Cystoscopy. A thin, lighted scope passed through the urethra reveals any abnormalities or cancerous tumors inside the bladder.

Voiding Cystourethrogram (VCUG). A special contrast liquid is placed inside the patient's bladder via a catheter, and x-ray pictures are taken while the bladder is filled and emptied. The images reveal defects in the urethra or bladder that could be allowing urine to back up into the kidneys.

Nuclear Scan. Radioactive liquid is injected into the patient's vein. As the liquid circulates through the urinary tract, it reveals whether the kidneys and bladder are functioning properly.

Treatment Options

Because they are bacterial in nature, nearly all UTIs are treatable with antibiotics (see Table 9.1). Which antibiotic the doctor chooses will depend on the results of the sensitivity test (discussed earlier in this chapter).

If the infection is not caused by a blockage or structural abnormality, a three-day course of oral antibiotics is usually enough to kill the bacteria, but doctors often recommend a full seven-day course to be on the safe side. Infections caused by chlamydia or mycoplasma are usually treated with a longer course of tetracycline, trimethoprim/sulfamethoxazole (TMP/SMZ), or doxycycline. While the antibiotics are taking effect, patients are advised

TABLE 9.1. Antibiotics Used to Treat UTIs

Antibiotic	Brand Name
Trimethoprim	Trimpex
Sulfamethoxazole	Bactrim
Amoxicillin	Amoxil
Nitrofurantoin	Macrodantin
Ofloxacin	Floxin
Norfloxacin	Noroxin
Ciprofloxacin	Cipro—also used to treat anthrax
Trovafloxin	Trovan

to use a heating pad to relieve any abdominal pain or pressure, and to drink plenty of water to cleanse their urinary tract of bacteria. After patients have completed the full course of antibiotics, their doctor will do a follow-up exam to make sure the infection has been eliminated.

Just because an infection has been wiped out with antibiotics does not mean that it is gone for good. Nearly 20 percent of women who have had one UTI can expect to have another at some point in their lives. The more infections a woman has had, the greater the likelihood that she will experience another. After three UTIs, 80 percent of women will have at least one more recurrence. Interestingly, the subsequent infections are usually caused by a different bacterial strain than the preceding infection. Doctors must test the patient's urine with each occurrence to isolate the correct bacteria and target the proper antibiotic treatment. Recurrent infections may require longer doses of antibiotics. In more severe cases, the patient may need to receive a dose of intravenous antibiotics in the hospital.

Rarely will a UTI result in long-term complications, but if left untreated, infections have been known to cause kidney stones (see Chapter 4) or scarring that eventually results in long-term kidney failure (however, serious damage is usually confined to people who have diabetes or another condition that predisposes them to kidney damage). Kidney damage is especially of concern among young children, because it can interfere with a child's growth and physical development.

Preventing Urinary Tract Infections

Women who have three or more UTIs a year may benefit from a daily dose of antibiotics over a period of six months or more. Alternately, doctors may prescribe a more selective course of antibiotics, which are taken

How to Prevent a UTI Recurrence

- Urinate regularly every 2–3 hours and always empty the bladder completely.
- After a bowel movement, women should wipe from front to back to avoid spreading bacteria from the rectal opening into the vagina and urethra.
- Avoid baths, especially bubble baths.
- Wash the genitals before and after intercourse. Women can also urinate after intercourse to flush out any bacteria that might have entered from the man's penis.
- Avoid feminine hygiene sprays and scented douches, which can irritate the vagina.
- Drink at least eight to ten glasses of water a day.
- Avoid such bladder irritants as coffee, alcohol, caffeine, and spicy foods, especially during an infection.

only after intercourse or after the symptoms of an infection appear. Women who prefer a more natural approach might try a glass a day of cranberry juice, which has been proven in studies to prevent against UTIs by inhibiting the growth of bacteria on the bladder wall. (See "How to Prevent a UTI Recurrence," and "A Glass of Cranberry Juice a Day May Keep the Urinary Tract Infections Away.") Daily vitamin C supplements appear to offer a similar benefit.

Urinary Tract Infections in Children

Aside from bedwetting, UTIs are the most common urinary problem in children (see Chapter 8). According to the National Institute of Diabetes and Digestive and Kidney Diseases ("Urinary Tract Infections in Children," 2000), an estimated 3 percent of girls and 1 percent of boys will have a UTI by the time they reach age 11. UTIs are particularly dangerous in children, because they can have a long-term effect on kidney growth and function.

Girls are more susceptible to infection than boys for the same reason women are more susceptible than men—their short urethra provides easy access for bacteria to enter the bladder. Babies in general are at risk for an infection because they sit in soiled diapers, which allows *E. coli* from fecal matter easy access to the urethra. In many cases, children develop recurrent UTIs because they were born with a urinary tract defect that allows urine to stagnate and breed bacteria. The most common abnormality is vesi-

A Glass of Cranberry Juice a Day May Keep the Urinary Tract Infections Away

Decades ago, people began to realize that cranberry juice was more than a tasty beverage; when taken regularly, it protected against bacterial bladder infections. In recent years, a number of scientific studies have proved the effectiveness of this heretofore home remedy. One such study, published in the *British Medical Journal* (Kontiokari et al., 2001) tested a cranberry-lingonberry juice concentrate on a group of 150 women, each of whom had an active urinary tract infection. The study found that women who drank cranberry juice every day for six months were 20 percent less likely to have a recurrence than women in the control group.

Why the protective benefit? Cranberries contain substances (one of which is the sugar, fructose) that prevent harmful *E. coli* bacteria (the leading cause of urinary tract infections) from colonizing the cells lining the bladder. Scientists say cranberry juice is not a replacement for antibiotics, but when taken regularly, it can reduce the incidence of infection and thus the need for antibiotic treatments in high-risk individuals.

coureteral reflux, a condition in which urine backs up from the bladder toward the kidneys. An overly narrow ureter or urethra, a kidney stone, or any other impediment to urine flow can also result in an infection.

The symptoms of an infection are easy to miss in very young children, who are unable to communicate pain and discomfort. Doctors advise parents to look out for some of the most common signs:

- Irritability
- Loss of appetite
- Unexplained fever that does not go away
- Strange-smelling or cloudy urine
- Crying during urination
- Inability to produce more than a few drops of urine
- Frequent urination
- Loose bowels
- Incontinence in children who are already toilet trained

Doctors diagnose a UTI in children by taking a urine sample, just as they do with adults. Like adults, older children can simply urinate into a sterile cup. But urine collection in children who are not yet toilet trained is more difficult. The doctor can either place a collection bag over the child's genitals and seal it with adhesive tape, or insert a catheter into the child's bladder to drain the urine. The urine is then examined under a microscope and cultured in a laboratory to determine which antibiotic will be most effective in killing the particular strain of bacteria causing the infection. If the urine culture is inconclusive, the doctor may call for one of the imaging tests mentioned earlier in this chapter.

Children must be monitored more closely than adults because they are at greater risk for permanent kidney damage. But serious complications are rare, and infections usually clear up with a 3–5-day course of antibiotics. After a child has finished the full dose of antibiotics, the doctor will schedule a follow-up visit to make sure that the infection is gone. If the infection stemmed from a congenital abnormality such as VUR, the doctor may recommend surgery to correct the problem.

Future Directions in the Treatment of Urinary Tract Infections

In the late 1970s, researchers discovered that a protective antibody, called Immunoglobulin A, was missing from the vaginal secretions of women who are prone to recurrent UTIs. Currently, the only way for these women to prevent future infections is with prophylactic doses of antibiotics. But doctors are concerned that bacteria can become resistant to antibiotics over time, rendering the drugs useless. So researchers are working to develop vaccines that trigger the body's immune system to fight off infection. The vaccines under investigation may be administered orally or via an injection, or released into the vagina via a suppository. In recent trials, suppositories have effectively protected women from infection for up to six months.

Acronyms

ADH Antidiuretic hormone

ADPKD Autosomal dominant polycystic kidney disease

AMACR A-methylacyl-CoA racemase

ARPKD Autosomal recessive polycystic kidney disease

ATP Adenosine triphosphate

BCG Bacillus Calmette-Guerin solution

BPH Benign prostatic hyperplasia

BTA Bladder tumor antigen

BUN Blood urea nitrogen

CAPD Continuous ambulatory peritoneal dialysis

CCPD Continuous cycling peritoneal dialysis

CT Computed tomography

DCT Distal convoluted tubule

DDAVP Desmopressin

DHT Dihydrotestosterone

DMSO Dimethyl sulfoxide

DRE Digital rectal exam

EKG Electrocardiogram

EMG Electromyograph

EPO Erythropoietin

ERT Estrogen replacement therapy

ESRD End-stage renal disease

ESWL Extracorporeal shock wave lithotripsy

FDA U.S. Food and Drug Administration

FSGS Focal segmental glomerulosclerosis

GFR Glomerular filtration rate

HLA Human leukocyte antigen

IC Interstitial cystitis

IMRT Intensity modulated radiation therapy

IPD Intermittent peritoneal dialysis

IVP Intravenous pyelogram

LHRH Luteinizing hormone-releasing hormone agonists

MCD Minimal change disease

MRI Magnetic resonance imaging

NIPD Nocturnal intermittent peritoneal dialysis

NMP Nuclear matrix protein

NSAIDs Nonsteroidal anti-inflammatory drugs

PCT Proximal convoluted tubule

PET Peritoneal equilibration test

PKD Polycystic kidney disease

PRA Panel reactive antibody

PSA Prostate specific antigen

PTH Parathyroid hormone

RNASEL Ribonuclease L

RPP Radical perineal prostatectomy

RRP Radical retropubic prostatectomy

RTA Renal tubular acidosis

SIC Self-intermittent catheterization

STD Sexually transmitted disease

TENS Transcutaneous electrical nerve stimulation

TUEP Transurethral evaporation of the prostate

TUIP Transurethral incision of the prostate

TUMT Transurethral microwave thermotherapy

TUNA Transurethral needle ablation

TUR Transurethral resection

TURP Transurethral resection of the prostate

TVP Transurethral electrovaporization of the prostate

TVT Tension-free vaginal tape

UNOS United Network for Organ Sharing

URR Urea reduction ration

UTI Urinary tract infection

VCUG Voiding cystourethrogram

VLAP Noncontact visual laser ablation

VUR Vesicoureteral reflux

Glossary

5 alpha-reductase inhibitors Drugs that shrink the prostate by blocking conversion of testosterone to DHT.

ACE inhibitors A class of drugs that lower blood pressure and reduce proteinuria.

Acidosis An abnormal increase in the acidity of the body's fluids.

Adenosine triphosphate (ATP) Energy used by cells in active transport.

Adipose capsule The central layer surrounding the kidney, composed of fatty tissue.

Adrenal androgen inhibitors Drugs that treat prostate cancer by blocking production of androgens by the adrenal gland.

Aldosterone A hormone secreted by the adrenal glands in the kidneys that increases sodium reabsorption.

Alpha adrenergic blockers Drugs that relax the smooth muscles of the prostate and bladder neck to relieve prostatic obstruction.

Amyloidosis Abnormal protein deposits in the kidneys.

Androgens Male hormones.

Antiandrogen agents Drugs that treat prostate cancer by blocking the body's ability to use androgens produced by the adrenal gland.

Antidiuretic hormone (ADH) A hormone secreted by the hypothalamus that controls permeability of the cells around the distal convoluted tubule and collecting duct of the nephrons.

Benign prostatic hyperplasia (BPH) (benign prostatic hypertrophy) Noncancerous growth of the prostate, which usually begins when a man reaches middle age.

Bladder Hollow muscular organ that stores urine for elimination.

Bladder exstrophy Congenital abnormality in which the bladder is turned inside out and protrudes through a hole in the abdomen.

Bladder washing Test in which doctors place a saline solution in the bladder, then remove it to test for cancerous cells.

Bowman's capsule Cup-shaped sac that surrounds the glomeruli of the nephrons in the kidneys. Also called glomerular capsule.

Brachytherapy Radioactivity delivered internally to the site of a cancerous tumor.

Calcitonin A hormone produced by the thyroid gland that decreases the amount of calcium excreted by the kidneys.

Calcitrol A hormone secreted by the kidneys that increases the levels of calcium and phosphorous in the blood.

Calculi Another name for stones, which are formed by the buildup of chemical crystals in urine.

Catheter A flexible tube used to inject materials into the body or to remove urine in patients who are unable to urinate naturally.

Citrate A form of citric acid contained in the urine that blocks certain types of kidney stone formation.

Clearance study A urine test used to determine how well the kidneys are filtering certain substances, including protein, creatinine, and urea.

Collecting duct Where fluid is carried from the distal convoluted tubule (DCT) in the nephron of the kidneys on its way to the minor calyx.

Computed tomography (CT) A procedure that uses a series of x-ray beams to create a computerized image of an organ.

Cortex Outer portion of the kidney.

Creatinine Waste produced by the breakdown of creatine phosphate in muscles.

Cystectomy Surgery to remove the bladder.

Cystinuria An inherited metabolic disorder in which the kidneys excrete too much of the amino acid cystine into the urine.

Cystitis A bacterial infection of the bladder.

Cystometry A test for evaluating incontinence that measures bladder pressure as well as maximum bladder capacity.

Cystoscope A thin instrument used to view the urinary tract.

Cystoscopy Test that involves insertion of a thin scope through the urethra into the bladder to check the bladder for tumors, stones, and ulcers and to measure bladder capacity.

Cytology Test in which doctors examine the patient's urine under a microscope to screen for cancerous cells.

Detrusor muscle Three layers of smooth muscle surrounding the mucosa of the bladder.

Dialysis An artificial filtration process that cleanses the blood in place of nonfunctioning kidneys.

Dialyzer Artificial kidney in which blood is filtered during hemodialysis.

Diffusion A passive process in which molecules move from an area of lower concentration to an area of higher concentration until a balance is reached.

Dihydrotestosterone (DHT) A testosterone derivative formed in the prostate.

Distal convoluted tubule (DCT) Located between the loop of Henle and the collecting duct inside the nephron of the kidney.

Diuretic A substance (i.e., caffeine) that increases urine production.

Dysuria Pain during urination.

Edema The accumulation of excess fluids in the tissues, which can result in swelling in the face, ankles, or wrists.

Electrofulguration A procedure that burns away cancer cells with high-energy electricity.

Electrolytes Chemical compounds (which include sodium, potassium, calcium, and chloride) that take on an electric charge when mixed into a solution.

Endoscopy The use of a thin instrument to visualize the interior of the urinary tract.

Epithelium Layer of cells that covers most of the body and its organs.

Erythropoietin A hormone secreted by the kidneys that stimulates red blood cell production.

External urethral sphincter Ring of voluntary muscle surrounding the end of the urethra, which regulates urine flow out of the body.

Extracellular fluid Fluid located outside of the cells; composed of interstitial fluid and blood plasma.

Glomerular capsule Cup-shaped sac that surrounds glomeruli of the nephrons in the kidneys; also called Bowman's capsule.

Glomerular filtrate The product of blood filtration in the nephrons of the kidneys.

Glomerulations Pinpoint bleeding on the bladder walls in patients with interstitial cystitis.

Glomerulonephritis A disease marked by inflammation of the glomeruli, or filtering units, of the kidneys.

Glomerulosclerosis A scarring or hardening of the glomeruli.

Glomerulus A network of tiny blood vessels in the nephron.

Glycosuria A condition characterized by the presence of glucose, or sugar, in the urine.

Gout A disorder, marked by high levels of uric acid in the blood, which causes painful joint inflammation.

Gross hematuria Blood in the urine that can be seen with the naked eye.

Hematuria Blood in the urine.

Hemodialysis An artificial kidney that filters wastes and excess fluids out of the blood and regulates chemical balance in place of damaged kidneys.

Hilius Curved notch on the side of each kidney near the center where blood vessels enter and exit the kidney.

Homeostasis Maintaining a balance of fluids within the body.

Hypercalciuria An excess of calcium in the urine, which can lead to calcium stone formation.

Hypernatremia Too much sodium in the extracellular fluid.

Hyperoxaluria An excess of oxalate in the urine, which can lead to calcium oxalate stone formation.

Hyperuricuria An excess of uric acid in the urine, which can lead to uric acid stone formation.

Hyponatremia Too little sodium in the extracellular fluid.

Hypothalamus Part of the brain that regulates fluid and electrolyte balance and hormone secretion.

Ileal conduit Procedure used in patients who have had their bladder removed, in which a portion of the small intestine is used as a conduit for urine to be removed through an opening in the abdomen.

Immunotherapy A cancer therapy that teaches the body's immune system to kill cancer cells.

Incontinence Inability to control the micturition, or urination, process.

Internal urethral sphincter Ring of involuntary muscle that surrounds the urethra where it meets the bladder and that controls the flow of urine.

Interstitial fluid The fluid contained in the spaces between cells.

Intracellular fluid Fluid that is contained within the body's cells.

Intravenous pyelogram (IVP) or intravenous urography (IVU) An x-ray technique that uses a special contrast dye to allow doctors to view an image of the urinary tract.

Ketone bodies A byproduct of fatty acid breakdown.

Kidneys The two bean-shaped organs that filter wastes, regulate electrolyte balance, and secrete hormones.

Lithotripsy A technique of breaking up or crushing urinary stones for removal.

Loop of Henle U-shaped section between the proximal convoluted tubule and the distal convoluted tubule in the nephron of the kidney.

Luteinizing hormone-releasing hormone (LHRH) agonists Drugs that treat prostate cancer by decreasing testosterone production in the testicles.

Lymphadenectomy Removal of the lymph nodes to determine whether cancer has spread.

Magnetic resonance imaging (MRI) A combination of magnetic field and radio waves used to create cross-sectional images of organs and tissues.

Major calyx Openings in the center of the kidneys through which urine flows into the renal pelvis.

Medulla Inner region of the kidney.

Microalbuminuria Leakage of the protein albumin into the urine from damaged glomeruli.

Microscopic hematuria Blood in the urine that is too small to be seen with the naked eye.

Micturition The process in which urine is released from the bladder; urination.

Minor calyx Cup-like receptacle attached to each renal pyramid in the kidney.

Mucosa A mucous membrane that lines a body cavity.

Nephrectomy Surgery to remove part or all of a diseased kidney.

Nephrolithotomy Stone removal surgery.

Nephrologist A doctor who specializes in diseases of the kidneys.

Nephron Tiny capsules in the kidneys that filter wastes out of the blood and reabsorb water and necessary molecules back into the blood.

Nitrogenous wastes The byproduct of protein metabolism.

Nocturia Frequent urination at night.

Nuclear scan Test for urinary system abnormalities. Radioactive material is injected into the patient's vein, which reveals the shape and function of the urinary tract as well as any abnormalities.

Orchiectomy Prostate cancer treatment that involves the surgical removal of the testicles to halt testosterone production.

Osmoreceptors Neurons that warn the hypothalamus of deficient fluids in the body.

Osmosis Movement of water molecules through a selectively permeable membrane from a high concentration to a lower concentration.

Papillary duct Tube that drains urine from collecting ducts in the nephron and empties it into the minor calyx.

Parasympathetic nerves Under the direction of the parasympathetic nervous system, these nerves release the neurotransmitter acetylcholine, which causes the bladder to contract during urination.

Parathyroid glands Four kidney-shaped glands located near or within the thyroid gland that secrete parathroid hormones, which regulates calcium and phosphorous balance.

Parathyroid hormone Hormone released by the parathyroid glands that regulates calcium and phosphorous balance.

Perineum (perineal) The area that lies between the scrotum and anus in men, and the vagina and anus in women.

Peritoneal dialysis A form of dialysis in which wastes are filtered within the patient's abdominal cavity.

Pessary A device placed in the vagina to support the bladder.

pH Measures the acidity or alkalinity of body fluids.

Prostate Gland surrounding the top of the urethra in men that contributes nutrients to the seminal fluid.

Prostatectomy Removal of part or all of the prostate to treat prostate cancer.

Prostatitis An inflammation (sometimes bacterial) of the prostate.

Proteinuria Abnormal amounts of protein in the urine.

Proximal convoluted tubule (PCT) Tiny tubes in the nephrons of the kidneys through which glomerular filtrate passes and substances necessary to the body (i.e., water, sodium, and calcium) are reabsorbed into the bloodstream.

Pyelonephritis A bacterial infection of the kidneys.

Renal Relating to the kidneys.

Renal capsule Layer of fibrous membrane that forms the innermost layer of the kidney.

Renal fascia Outermost layer of the kidney, composed of connective tissue that holds the kidney to the abdominal wall.

Renal pelvis Funnel-shaped cavity that collects urine and sends it into the ureter.

Renal pyramids Cone-shaped receptacles inside the medulla of the kidney.

Renin An enzyme secreted by the kidneys that leads to the production of the hormone aldosterone.

Resectoscope A long, thin, lighted instrument (much like a cystoscope) used to perform transurethral resection of the prostate (TURP).

Semipermeable (or selectively permeable) membrane Membrane that allows certain molecules to pass through while restricting others.

Serosa Outer layer of the bladder wall.

Sodium/potassium pump A form of active transport that regulates the amount of sodium and potassium in and around the cells.

Sphincter electromyography An evaluation for incontinence that measures the strength of each sphincter muscle contraction during bladder filling and emptying.

Sphincters Two groups of muscles that control the flow of urine from the bladder.

Stoma An opening in the abdominal wall through which urine is diverted in patients who have had their bladder removed.

Synchronous multichannel videourodynamic studies A device used to diagnose urinary incontinence that measures bladder and rectum pressure as well as urinary flow rate.

Transurethral electrovaporization of prostate (TVP) A new version of TURP, which uses a high-frequency electrical current to cut and remove excess prostate tissue.

Transurethral evaporation of prostate (TUEP) This treatment for benign prostatic hyperplasia uses laser energy to heat and destroy prostate tissue.

Transurethral incision of the prostate (TUIP) A surgical procedure used to treat benign prostatic hyperplasia, in which two small cuts are made in the prostate to relieve pressure on the urethra.

Transurethral microwave thermotherapy (TUMT) A procedure used in the treatment of benign prostatic hyperplasia that uses microwave energy to heat and destroy obstructing prostate tissue.

Transurethral needle ablation (TUNA) Radio-frequency heat applied to the prostate via tiny needles to burn away obstructing tissue.

Transurethral resection (TUR) Procedure used to remove tumors on the lining of the bladder or prostate.

Transurethral resection of the prostate (TURP) A surgical procedure used to treat benign prostatic hyperplasia and prostate cancer by removing the inner tissue of the prostate.

Trigone A triangular-shaped region located in the bladder floor.

Ultrasound Uses high-frequency sound waves to produce an image of the kidneys, bladder, or other organs on a video screen.

Urea Waste produced by the breakdown of proteins.

Ureteral orifices Two holes where the ureters pierce the bladder.

Ureters Two thin, hollow tubes that carry urine from the kidneys to the urinary bladder.

Urethra Muscular tube that connects the bladder with the exterior of the body.

Urethritis A bacterial infection of the urethra.

Uric acid Waste produced by the breakdown of nucleic acids (DNA and RNA).

Urinalysis Test that detects the presence of certain cells or chemicals in the urine.

Urochrome Pigment produced by the breakdown of bile that gives urine its yellow or amber color.

Urodynamic evaluation Series of tests that determine a patient's type of urinary incontinence by gauging how well his or her bladder, urethra, and rectum are holding and releasing urine.

Uroflometry A test for incontinence that measures urinary flow rate to determine how well the bladder and urethra are working.

Uroscopy Examining a patient's urine to diagnose disease.

Vesicoureteral reflux A condition in which urine backs up from the bladder toward the kidneys, where it can stagnate and cause infection.

Voiding cystourethrogram A test to reveal abnormalities of the urethra, bladder, and urine flow, in which a special contrast liquid is placed inside the patient's bladder and x-ray pictures are taken while the bladder is filled and emptied.

Xenotransplantation Implantation of living cells or organs from another species.

Organizations and Web Sites

American Association of Kidney Patients
3505 E. Frontage Road, Suite 315
Tampa, FL 33607
Phone: (800) 749-2257
Email: info@aakp.org
http://www.aakp.org

This organization is designed to help kidney patients and their families deal with the physical, emotional, and social ramifications of their disease. The site includes a directory of local and national resources and a free monthly newsletter.

American Cancer Society
1599 Clifton Road, NE
Atlanta, GA 30329
Phone: (800) ACS-2345
http://www.cancer.org

The country's foremost cancer research and advocacy organization offers a wealth of background, treatment, and research information on every type of cancer.

American Diabetes Association
1701 North Beauregard Street
Alexandria, VA 22311
Phone: (800) DIABETES (342-2383)
http://www.diabetes.org

This site is designed for diabetics and their families. It includes background information on the disease and a guide to healthy living for diabetics.

American Foundation for Urologic Disease
1128 North Charles Street
Baltimore, MD 21201

Phone: (800) 242-2383
http://www.afud.org

The foundation has designed its Web site to keep patients and their families apprised of the latest developments in urologic disease research and treatments.

American Kidney Fund
6110 Executive Boulevard, Suite 1010
Rockville, MD 20852
Phone: (800) 638-8299
http://www.akfinc.org

The American Kidney Fund provides financial support as well as research and educational programs to patients with kidney disease.

American Society of Nephrology
2025 M Street NW, Suite 800
Washington, DC 20036
Phone: (202) 367-2190
http://www.asn-online.org

Although the ASN has geared its site to medical professionals, it contains statistics and other information on kidney disease that patients and their families might find useful.

International Society for Peritoneal Dialysis
http://www.ispd.org

The ISPD Web site provides up-to-date and relevant information on peritoneal dialysis research and breakthroughs.

Interstitial Cystitis Association
110 North Washington Street, Suite 340
Rockville, MD 20850
Phone: (800) HELP-ICA
http://www.ichelp.org

With a number of patient resources and a quarterly newsletter, this organization strives to educate the public about this little-known affliction.

Kidney Cancer Association
1234 Sherman Avenue, Suite 203
Evanston, IL 60202-1375
Phone: (800) 850-9132
http://www.kidneycancerassociation.org

An organization composed of kidney cancer patients and their families, which is devoted to advocacy and outreach programs.

National Association for Continence
P.O. Box 8306
Spartanburg, SC 29305
Phone: (800) BLADDER (252-3337)
http://www.nafc.org/site2/index.html

Offers information on the causes, prevention, diagnosis, and treatment options for incontinence.

National Cancer Institute
9000 Rockville Pike
Bethesda, MD 20892
Phone: (800) 4-CANCER
http://www.nci.nih.gov

Background information for patients and their families on a number of different cancers, as well as a listing of ongoing clinical trials.

National Kidney Foundation
30 East 33rd Street, Suite 1100
New York, NY 10016
Phone: (800) 622-9010
http://www.kidney.org

Offers an A-to-Z guide of kidney disease information for patients and their families.

The NephCure Foundation
Phone: (866) NEPHCURE (637-4287)
http://www.nephcure.org

A coalition of scientists, patients, and families, which addresses the problems and concerns related to kidney filtration diseases.

Polycystic Kidney Disease Foundation
4901 Main Street, Suite 200
Kansas City, MO 64112-2634
Phone: (800) PKD-CURE
http://www.pkdcure.org

This organization is devoted to promoting research, fostering awareness, and finding a cure for polycystic kidney disease.

Simon Foundation for Continence
P.O. Box 835-F
Wilmette, IL 60091
Phone: (800) 23SIMON
http://www.simonfoundation.org

The Simon Foundation helps patients cope with incontinence by providing support, education, and up-to-date treatment information.

United Network for Organ Sharing
1100 Boulders Parkway, Suite 500
Richmond, VA 23225-8770
http://www.unos.org

This national list keeps track of all patients waiting for organs, and matches them to available donors.

Bibliography

Ahlstrom, Timothy P. *The Kidney Patient's Book.* Delran, NJ: Great Issues Press, 1991.

American Academy of Family Physicians. "Interstitial Cystitis: A Bladder Problem." *American Family Physician* 64 (2001): 1212.

American Cancer Society. "Cancer Facts & Figures 2002." http://www.cancer.org/downloads/STT/CancerFacts&Figures2002TM.pdf.

American Cancer Society. "What Is Bladder Cancer?" http://www.cancer.org. June 2003. Accessed December 17, 2003.

American Diabetes Association. "Kidney Disease: Options for Prevention and Treatment." http://www.diabetes.org. Accessed December 18, 2001.

American Foundation for Urologic Disease. "Bladder Cancer." http://www.afud.org. Accessed February 24, 2002.

American Foundation for Urologic Disease. "Kidney Stones." http://www.afud.org. Accessed December 18, 2001.

American Foundation for Urologic Disease. "Prostate Disease: Vital Information for Men over 40." http://www.afud.org/conditions/Postatrdiscase.asp. Accessed March 18, 2002.

American Foundation for Urologic Disease. "What Are the Symptoms of Prostatitis?" http://www.afud.org/conditions/psduring.html. Accessed March 18, 2002.

American Foundation for Urologic Disease. "What Is BPH?" http://www.afud.org/conditions/bphduring.html. Accessed March 18, 2002.

American Foundation for Urologic Disease. "What Is a Urinary Tract Infection?" http://www.afud.org/conditions/utibefore.html. Accessed April 3, 2002.

American Society of Nephrology. "Kidney Disease Facts & Statistics." http://www.asn-online.org/facts/. Accessed December 18, 2001.

American Urological Association. "Prostate-Specific Antigen (PSA) Best Practice Policy." *Oncology* 14 (2000): 267–286.

Anderson, J., T. Hanna, and P. Fanti. "Soy Protein and Protection from Diabetic Kidney Disease." *The Journal of Nutrition* 130 (2000): 675.

Baerheim, A. "Empirical Treatment of Uncomplicated Cystitis." *British Medical Journal* 323 (2001): 1197–1198.

Barry, H. C., M. C. Ebell, and J. Hickner. "Evaluation of Suspected Urinary Tract Infection in Ambulatory Women: A Cost-Utility Analysis of Office-Based Strategies." *The Journal of Family Practice* 44 (1997): 49–60.

Bartlett, Stephen T., and Eugene J. Schweitzer. "Laparoscopic Living Donor Nephrectomy for Kidney Transplantation." *Dialysis & Transplantation* 28 (1999): 318–331.

Baskin, Leland B., and Rebecca Hsu. "Laboratory Evaluation of Proteinuria." *Medical Laboratory Observer* 31 (1999): 30.

Bastian, Glenn F. *An Illustrated Review of the Urinary System*. New York: HarperCollins College Publishers, 1994.

Bihl, Geoffrey, and Anthony Meyers. "Recurrent Renal Stone Disease—Advances in Pathogenesis and Clinical Management." *The Lancet* 358 (2001): 651.

Blaivas, Jerry G. *Conquering Bladder and Prostate Problems*. New York and London: Plenum Trade, 1998.

Bliwise, Donald L., et al. "Survival by Time of Day of Hemodialysis in an Elderly Cohort." *Journal of the American Medical Association* 286 (2001): 2690–2694.

Bochner, Bernard H., et al. "The Promise of Antiangiogenic Therapies." *Contemporary Urology* 13 (2001): 51.

Borghi, Loris, et al. "Comparison of Two Diets for the Prevention of Recurrent Stones in Idiopathic Hypercalciuria." *The New England Journal of Medicine* 346 (2002): 77–84.

Bowden, Rodney G., et al. "Homocysteine, Age, Gender and Vascular Access Thrombosis in End-Stage Renal Disease Patients: Retrospective Analysis." *Journal of Nephrology* 6 (2002): 666–67.

Brenner, Barry M., et al. "Effects of Losartan on Renal and Cardiovascular Outcomes in Patients with Type 2 Diabetes and Nephropathy." *The New England Journal of Medicine* 345 (2001): 861–869.

Brooks, James D., et al. "Plasma Selenium Level before Diagnosis and the Risk of Prostate Cancer Development." *The Journal of Urology* 166 (2001): 2034–2038.

Cancer Facts. "Kidney Cancer Overview." http://www.cancerfacts.com. Accessed December 18, 2001.

Carpten, John, et al. "Germline Mutations in the Ribonuclease L (RNASEL) Gene in Hereditary Prostate Cancer 1 (HPC1) Linked Families." *Nature Genetics* 30 (2002): 1–4.

Carroll, Michael F., and Jonathan L. Temte. "Proteinuria in Adults: A Diagnostic Approach." *American Family Physician* 62 (2000): 1333.

Chang, Kenneth. "Company Says It Used Cloning to Create New Kidneys for Cow." *The New York Times* (January 31, 2002). http://www.nytimes.com/2002/01/31/health/31CLON.html.

Charlton, C.A.C. *The Urological System*. Harmondsworth, UK: Penguin Books, 1973.

Childs, Richard, et al. "Regression of Metastatic Renal-Cell Carcinoma after Nonmyeloablative Allogeneic Peripheral-Blood Stem-Cell Transplantation." *The New England Journal of Medicine* 343 (2000): 750–758.

Chow, Wong-Ho, et al. "Obesity, Hypertension, and the Risk of Kidney Cancer in Men." *The New England Journal of Medicine* 343 (2000): 1305–1311.

"Clinical Trial Data Reported for Thalidomide Treatment of Kidney and Brain Cancers." *Angiogenesis Weekly* (June 22, 2001): 3.

Couser, William G. "Glomerulonephritis." *The Lancet* 353 (1999): 1509–1515.

Curhan, Gary C., et al. "Beverage Use and Risk for Kidney Stones in Women." *Annals of Internal Medicine* 128 (1998): 534–540.

Dahl, S. D. "William Cheselden (1688–1752)." *Invest. Urol.* 5 (1968): 627–629.

"Deaths: Preliminary Data for 2000." *National Vital Statistics Report* 49, no. 12 (October 9, 2001). National Center for Health Statistics. Bethesda, MD: Centers for Disease Control and Prevention.

DeBisschop, Michael E., and Catherine M. Oliphant. "Management of Urinary Incontinence." *U.S. Pharmacist* 27 (2002). http://www.uspharmacist.com/NewLook/DisplayArticle.cfm?item=_num=833.

Delzell, John E., and Michael L. Lefevre. "Urinary Tract Infections during Pregnancy." *American Family Physician* (February 1, 2000): 713–721.

Doering, Christopher. "Filtering Blood May Allow More Transplants: Study." *Reuters* (April 30, 2002).

Donadio, James V., et al. "A Controlled Trial of Fish Oil in IgA Nephropathy." *The New England Journal of Medicine* 331 (1994): 1194–1199.

"Drug Companies Start Phase II Monotherapy Clinical Trials." *Immunotherapy Weekly* (May 16, 2001).

Duley, Irene, and Patricia Gabow. *PKD Patient's Manual.* Kansas City, MO: The Polycystic Kidney Research Foundation, 1995.

Elliott, Victoria S. "New Bladder Cancer Tests May Aid Detection." *American Medical News* 43 (2000): 24.

Fickenscher, Lisa. "Evaluating Adult Hematuria." *The Nurse Practitioner* 24 (1999): 58.

Flanigan, Robert C., et al. "Nephrectomy Followed by Interferon Alfa-2b Compared with Interferon Alfa-2b Alone for Metastatic Renal-Cell Cancer." *The New England Journal of Medicine* 345 (2001): 1655–1659.

Fored, Michael C., et al. "Acetaminophen, Aspirin, and Chronic Renal Failure." *The New England Journal of Medicine* 345 (2001): 1801–1808.

Garell, Dale C., ed. *The Encyclopedia of Health: Kidney Disorders.* New York: Chelsea House Publishers, 1992.

Garely, Alan D. "Managing Urinary Incontinence: An Expanding Role for Ob/Gyns." http://www.obmanagement.com/content/obg=_featurexml.asp?file=2002/01/obg=_0102=_00026.xml. Accessed March 21, 2002.

"Genetically Engineered T Cell Tackles Tumors." *Vaccine Weekly* (December 12, 2001), 14.

Giovannucci, Edward, et al. "A Prospective Study of Tomato Products, Lycopene, and Prostate Cancer Risk." *Journal of the National Cancer Institute* 94 (2002): 391–398.

Gleich, Paul. "Hematuria: Just UTI—or Something More Ominous?" *Consultant* 39 (1999): 2235–2240.

Goldfarb, David S., and Fredric L. Coe. "Prevention of Recurrent Nephrolithiasis." *American Family Physician* 60 (1999): 2269.

Goldfarb, David S., et al. "Renal Outcome 25 Years after Donor Nephrectomy." *The Journal of Urology* 166 (2001): 2043–2047.

Grossfeld, Gary D., et al. "Asymptomatic Microscopic Hematuria in Adults: Summary of the AUA Best Practice Policy Recommendations (American Urological Association)." *American Family Practice* 63 (2001): 1145.

Gupta, K., D. F. Sahm, D. Mayfield, and W. E. Stamm. "Antimicrobial Resistance among Uropathogens that Cause Community-Acquired Urinary Tract Infection in Women: A Nationwide Analysis." *Clinical Infectious Diseases* 33 (2001): 89–94.

Harding, Anne. "Vaginal Vaccine Shows Promise in Bladder Infection." *Reuters Health* (December 18, 2001).

Hateboer, Nick, et al. "Comparison of Phenotypes of Polycystic Kidney Disease Types 1 and 2." *The Lancet* 353 (1999): 103–107.

Heaf, J.G., Hans Løkkegaard, and Melvin Madsen. "Initial Survival Advantage of Peritoneal Dialysis Relative to Haemodialysis." *Nephrology Dialysis Transplantation* 17 (2002): 112–117.

Henkel, John. "Prostate Cancer: No One Answer for Testing or Treatment." *FDA Consumer* (September–October 1998), FDA Publication No. 00-1297. http://www.fda.gov/fdoc/features/1998/598=_pros.html.

Herman, John R. *Urology: A View through the Retrospectroscope.* Hagerstown, MD: Harper & Row, 1973.

Hitt, Emma. "Vaccine Elicits Response in Patients with Advanced Prostate Cancer." *Reuters Health* (April 22, 2002).

Ho, E.T., et al. "The Haematuria Clinic—Referral Patterns in Northern Ireland." *Ulster Medical Journal* 67 (1998): 25–28.

Hock, L., James Lynch, and K.C. Balaji. "Increasing Incidence of All Stages of Kidney Cancer in the Last 2 Decades in the United States: An Analysis of Surveillance, Epidemiology and End Results Program Data." *The Journal of Urology* 167 (2002): 57–60.

Holmberg, Lars, et al. "A Randomized Trial Comparing Radical Prostatectomy with Watchful Waiting in Early Prostate Cancer." *The New England Journal of Medicine* 347 (2002): 781–789.

Hunt, Jennifer. *Irritable Bladder & Incontinence: A Natural Approach.* Berkeley, CA: Ulysses Press, 1998.

The International Society for Peritoneal Dialysis. "The Emergence of Peritoneal Dialysis." http://www.ispd.org/history/genesis.php3. Accessed December 18, 2001.

Interstitial Cystitis Association. "What Is Interstitial Cystitis?" http://www.ichelp.org/. Accessed March 2, 2002.

Kontiokari, T., et al. "Randomised Trial of Cranberry-Lingonberry Juice and Lactobacillus GG Drink for the Prevention of Urinary Tract Infections in Women." *British Medical Journal* 322 (2001): 1571–1573.

"Laparoscopic Kidney Removal Is Effective Treatment." *Cancer Weekly* (June 19, 2001).

Lau, Weber K.O., et al. "Matched Comparison of Radical Nephrectomy vs. Nephron-Sparing Surgery in Patients with Unilateral Renal Cell Carcinoma and a Normal Contralateral Kidney." *Mayo Clinic Proceedings* 75 (2000): 1236–1242.

Lee, H.S.J., ed. *Dates in Urology.* New York and London: Parthenon Publishing Group, 2000.

Lewis, Edmund J., et al. "Renoprotective Effect of the Angiotensin-Receptor Antagonist Irbesartan in Patients with Nephropathy Due to Type 2 Diabetes." *The New England Journal of Medicine* 345 (2001): 851–860.

March of Dimes. "Genital and Urinary Tract Defects." http://www.modimes.org/HealthLibrary2/FactSheets/GenitalAndUrinaryTract.htm. 2001. Accessed February 24, 2002.

Mariani, A.J., et al. "The Significance of Adult Hematuria: 1,000 Hematuria Evaluations Including a Risk-Benefit and Cost-Effectiveness Analysis." *Journal of Urology* 141 (1989): 350–355.

Mayo Foundation for Medical Education and Research. "What Is Prostate Cancer?"

http://www.mayoclinic.com/findinformation/diseasesandconditions/invoke.cfm?id=DS00043. October 9, 2000. Accessed March 18, 2002.

Mayo Foundation for Medical Education and Research. "What Is Prostate Gland Enlargement?" http://www.mayoclinic.com/findinformation/diseasesandconditions/invoke.cfm?id=DS00027. March 1, 2002. Accessed March 18, 2002.

Mickisch, G.H.J., et al. "Radical Nephrectomy Plus Interferon-Alfa-Based Immunotherapy Compared with Interferon Alfa Alone in Metastatic Renal-Cell Carcinoma: A Randomised Trial." *The Lancet* 358 (2001): 966–970.

Miller, Barry A., et al. *Racial/Ethnic Patterns of Cancer in the United States 1988–1992*. NIH Pub. No. 96–4104. Bethesda, MD: National Cancer Institute, 1996.

Murphy, Leonard J. *The History of Urology*. Springfield, IL: Charles C. Thomas, 1972.

National Association for Continence. "What Is Incontinence and What Causes It?" http://www.nafc.org. Accessed March 21, 2002.

National Cancer Institute. "Bladder Cancer." http://cancernet.nci.nih.gov/wyntk/bladder. Updated January 22, 2002. Accessed February 24, 2002.

National Cancer Institute. "Kidney Cancer." http://cancernet.nci.nih.gov/wyntk/kidney. September 28, 1998. Updated December 12, 2000. Accessed December 18, 2001.

National Cancer Institute. "Pre-Surgical Chemotherapy May Improve Survival in Bladder Cancer." http://cancertrials.nci.nih.gov/news/meetings/asco01/blad051401.html. May 14, 2001.

National Cancer Institute. "Questions and Answers about the Prostate-Specific Antigen (PSA) Test." http://cis.nci.nih.gov/fact/5=_29.html. Updated January 11, 2001. Accessed March 18, 2002.

National Cancer Institute. "Transitional Cell Cancer of the Renal Pelvis and Ureter." http://www.cancernet.nci.nih.gov. Updated December 2001. Accessed December 18, 2001.

National Cancer Institute. "What You Need to Know about Prostate Cancer." NIH Publication No. 00-1576. http://www.cancer.gov/cancer=_information/doc=_wyntk.aspx?viewid=b94a9092-bbc1-4ba2-8c75-6793238d92a4. December 5, 2000. Updated January 22, 2002. Accessed March 18, 2002.

National Cancer Institute. "Wilms' Tumor and Other Childhood Kidney Tumors." http://www.cancernet.nci.nih.gov/cancer=_types/wilms=_tumor.shtml. December 2001. Accessed December 18, 2001.

National Diabetes Information Clearinghouse. "Diabetes Control and Complications Trial (DCCT)." http://www.niddk.nih.gov/health/diabetes/pubs/dcct1/dcct.htm. October 2001. Accessed December 18, 2001.

National Diabetes Information Clearinghouse. "Kidney Disease of Diabetes." http://www.niddk.nih.gov/health/diabetes/pubs/kdd/kdd.htm. February 12, 1998. Accessed December 18, 2001.

National Institute of Allergy and Infectious Diseases. "Fact Sheet: Wegener's Granulomatosis." http://www.niaid.gov/factsheets/wegeners.htm. May 1997. Accessed December 19, 2001.

National Institute of Diabetes and Digestive and Kidney Diseases. "Amyloidosis and Kidney Disease." http://www.niddk.nih.gov/health/kidney/pubs/kidney.failure/amyloidosis/amyloidosis.htm. May 2001. Accessed December 18, 2001.

National Institute of Diabetes and Digestive and Kidney Diseases. "Anemia in Kidney Disease and Dialysis." http://www.niddk.nih.gov/health/kidney/pubs/kidney-failure/anemia/anemia.htm. May 2001. Accessed February 4, 2002.

National Institute of Diabetes and Digestive and Kidney Diseases. "Cystocele (Fallen

Bladder)." http://www.niddk.nih.gov/health/urolog/summary/cystocel/index.htm. Updated May 1999. Accessed February 24, 2002.

National Institute of Diabetes and Digestive and Kidney Diseases. "Hemodialysis Dose and Adequacy." http://www.niddk.nih.gov/health/kidney/pubs/kidney-failure/hemodialysis-dose/hemodialysis-dose.htm. May 2001. Accessed December 18, 2001.

National Institute of Diabetes and Digestive and Kidney Diseases. "Kidney Failure: Choosing a Treatment that's Right for You." http://www.niddk.nih.gov/health/kidney/pubs/kidney-failure/choosing-a-treatment/choosing-a-treatment.htm. May 2001. Accessed December 18, 2001.

National Institute of Diabetes and Digestive and Kidney Diseases. "Neurogenic Bladder." http://www.niddk.nih.gov/health/urolog/summary/neuro/index.htm. September 1998. Accessed February 24, 2002.

National Institute of Diabetes and Digestive and Kidney Diseases. "Prevent Diabetes Problems: Keep Your Kidneys Healthy." http://www.niddk.nih.gov/health/diabetes/pubs/complications/kidneys/kidneys.htm. August 2000. Accessed December 18, 2001.

National Institute of Diabetes and Digestive and Kidney Diseases. "Prostate Enlargement: Benign Prostatic Hyperplasia." http://www.niddk.nih.gov/health/urolog/pubs/prostate/index.htm. May 1998. Updated January 2000. Accessed March 18, 2002.

National Institute of Diabetes and Digestive and Kidney Diseases. "Treatment Methods for Kidney Failure: Peritoneal Dialysis." http://www.niddk.nih.gov/health/kidney/pubs/kidney-failure/treatment-peritoneal/treatment-peritoneal.htm. May 2001. Accessed February 4, 2002.

National Institute of Diabetes and Digestive and Kidney Diseases. "Urinary Tract Infections in Children." http://Kidney.niddk.nih.gov/kudiseases/pubs/utichildren. September 2000. Accessed December 18, 2003.

National Institute of Diabetes and Digestive and Kidney Diseases. "UTI Vaccine on the Horizon." *Research Updates* (Summer 1999–2000): http://www.niddk.nih.gov/health/kidney/Research=_Updates/sum99/1.htm#a. Accessed March 21, 2002.

National Institute of Diabetes and Digestive and Kidney Diseases. "Vascular Access for Hemodialysis." http://www.niddk.nih.gov/health/kidney/pubs/kidney-failure/vascular-access/vascular-access.htm. May 2001. Accessed February 4, 2002.

National Institute of Diabetes and Digestive and Kidney Diseases. "Vesicoureteral Reflux." http://www.niddk.nih.gov/health/kidney/summary/vesico/vesico.htm. Updated March 2001. Accessed December 18, 2001.

National Institutes of Health. "Radiofrequency Energy Sizzles Tumors without Surgery." NIH News Release (November 26, 2001).

National Institutes of Health. "Two-Drug Therapy is Best for Symptomatic Prostate Enlargement." NIH News Release (May 28, 2002).

National Kidney and Urologic Diseases Information Clearinghouse. "Analgesic Nephropathy (Painkillers and the Kidneys)." http://www.niddk.nih.gov/health/kidney/summary/analgesc/index.htm. October 2, 1998. Accessed December 18, 2001.

National Kidney and Urologic Diseases Information Clearinghouse. "Childhood Nephrotic Syndrome." http://www.niddk.nih.gov/health/kidney/pubs/cns/cns.htm. June 2000. Accessed December 18, 2001.

National Kidney and Urologic Diseases Information Clearinghouse. "Glomerular

Diseases." http://www.niddk.nih.gov/health/kidney/pubs/glomer/glomer.htm. June 24, 1999. Accessed December 18, 2001.
National Kidney and Urologic Diseases Information Clearinghouse. "Goodpasture Syndrome." http://www.niddk.nih.gov/health/kidney/summary/good/index.htm. October 1998. Updated March 2001. Accessed December 18, 2001.
National Kidney and Urologic Diseases Information Clearinghouse. "Growth Failure in Children with Kidney Disease." http://www.niddk.nih.gov/health/kidney/summary/growfail/index.htm. Updated June 24, 1999. Accessed December 18, 2001.
National Kidney and Urologic Diseases Information Clearinghouse. "Hematuria (Blood in the Urine)." http://www.niddk.nih.gov/health/urolog/summary/hematuri/index.htm. May 1999. Accessed December 18, 2001.
National Kidney and Urologic Diseases Information Clearinghouse. "IgA Nephropathy." http://www.niddk.nih.gov/health/kidney/summary/iganeph/iganeph.htm. Updated May 2001. Accessed December 18, 2001.
National Kidney and Urologic Diseases Information Clearinghouse. "Interstitial Cystitis." http://www.niddk.nih.gov/health/urolog/pubs/cystitis/cystitis.htm. Updated February 2000. Accessed March 2, 2002.
National Kidney and Urologic Diseases Information Clearinghouse. "Kidney and Urologic Diseases Statistics for the United States." http://www.niddk.nih.gov/health/kidney/pubs/kustats/kustats.htm. December 2001. Accessed February 4, 2002.
National Kidney and Urologic Diseases Information Clearinghouse. "Kidney Stones in Adults." http://www.niddk.nih.gov/health/kidney/pubs/stonadul/stonadul.htm. February 2000. Accessed December 18, 2001.
National Kidney and Urologic Diseases Information Clearinghouse. "Nephrotic Syndrome in Adults." http://www.niddk.nih.gov/health/kidney/summary/nephsynd/nephsynd.htm. Accessed December 18, 2001.
National Kidney and Urologic Diseases Information Clearinghouse. "Polycystic Kidney Disease." www.niddk.nih.gov/health/kidney/pubs/polycyst/polycyst.htm. February 12, 1998. Accessed December 18, 2001.
National Kidney and Urologic Diseases Information Clearinghouse. "Proteinuria." http://www.niddk.nih.gov/health/kidney/pubs/proteinuria/proteinuria.html. December 2000. Accessed December 18, 2001.
National Kidney and Urologic Diseases Information Clearinghouse. "Renal Osteodystrophy." http://www.niddk.nih.gov/health/kidney/pubs/kidney-failure/renal-osteodystrophy/renal-osteodystrophy.htm. May 2001. Accessed December 18, 2001.
National Kidney and Urologic Diseases Information Clearinghouse. "Renal Tubular Acidosis." http://www.niddk.nih.gov/health/kidney/pubs/rta/rta.htm. June 2000. Accessed December 18, 2001.
National Kidney and Urologic Diseases Information Clearinghouse. "Urinary Incontinence in Children." NIH Publication No. 01-4095 (January 8, 2001).
National Kidney and Urologic Diseases Information Clearinghouse. "Urinary Incontinence in Women." http://www.niddk.nih.gov/health/urolog/pubs/uiwomen/uiwomen.htm. Updated February 11, 1998. Accessed March 21, 2002.
National Kidney and Urologic Diseases Information Clearinghouse. "Urinary Tract Infection in Adults." http://www.niddk.nih.gov/health/urolog/pubs/utiadult/utiadult.htm. Updated April 2001. Accessed March 24, 2002.
National Kidney and Urologic Diseases Information Clearinghouse. "Your Urinary

System and How It Works." http://www.niddk.nih.gov/health/urolog/pubs/yrurinar/index.htm. April 1998. Accessed November 15, 2001.
National Kidney Foundation, Inc. "Diabetes and Chronic Kidney Disease: Ten Facts." http://www.kidney.org/general/atoz/content/10factsdiabetes.html. Accessed December 18, 2001.
National Kidney Foundation, Inc. "End Stage Renal Disease in the United States." http://www.kidney.org/general/news/factsheet.cfm?id=38. November 2003. Accessed December 17, 2003.
National Kidney Foundation, Inc. "Getting Ready for a Transplant." http://www.kidney.org/general/atoz/content/transgettingready.html. Updated July 30, 2001. Accessed February 10, 2002.
National Kidney Foundation, Inc. "New Techniques for Treating Kidney Stones: Extracorporeal Shock Wave Lithotripsy." http://www.kidney.org/general/atoz/. Updated May 15, 2001. Accessed December 18, 2001.
National Kidney Foundation, Inc. "Nutrition and Changing Kidney Function." http://www.kidney.org/general/atoz/content/nutkidfunc.html. Updated May 15, 2001. Accessed December 18, 2001.
National Kidney Foundation, Inc. "The Problem of Kidney and Urologic Disease." http://www.kidney.org/general/news/factsheet.cfm?id=11. January 2003.
National Kidney Foundation, Inc. "The Transplant Waiting List." http://www.kidney.org/general/atoz/content/transwaitlist.html. Updated July 30, 2001. Accessed December 18, 2001.
The Nemours Foundation. "Kidney Diseases in Childhood." http://www.kidshealth.org/parent/medical/kidney/kidney=_diseases=_childhood.html. May 2001. Accessed December 18, 2001.
The Nemours Foundation. "When Your Child Has a Chronic Kidney Disease." http://www.kidshealth.org/parent/medical/kidney/kidney=_diseases=_childhood.html. May 2001. Accessed December 18, 2001.
Nitti, Victor W. "Strategies for Effective Evaluation and Management of Overactive Bladder." *Contemporary Urology* 13 (2001): 14.
Overmyer, Mac. "Antibiotic Therapy May Pose Risk for Stone Disease: May Eradicate Beneficial O Formigenes Bacteria, Leading to Stone Disease." *Urology Times* (July 1, 2001).
Oxalosis and Hyperoxaluria Foundation. "About Primary Hyperoxaluria." http://www.ohf.org/ph1.html. Accessed December 19, 2001.
Parfrey, P. S., and R. N. Foley. "The Clinical Epidemiology of Cardiac Disease in Chronic Renal Failure." *Journal of the American Society of Nephrology* 7 (1999): 1606–1615.
Parving, Hans-Henrik, et al. "The Effect of Irbesartan on the Development of Diabetic Nephropathy in Patients with Type 2 Diabetes." *The New England Journal of Medicine* 345 (2001): 870–878.
"Pig Donor 'Breakthrough' Claimed." Cnn.com. http://www.cnn.com/2002/WORLD/europe/08/23/uk.pigs/index.html. Accessed August 23, 2002. December 18, 2001.
Pirtskalaishvili, Giorgi, et al. "Update on Urine-Based Markers for Bladder Cancer." *Postgraduate Medicine* 106 (1999). http://www.postgradmed.com/issues/1999/11–99/Pirtskallaishvili.htm.
Polycystic Kidney Foundation. "Learning about Polycystic Kidney Disease." http://www.pkdcure.org/aboutpkd.htm. Accessed December 20, 2001.

Portis, Andrew J., and Chandru P. Sundaram. "Diagnosis and Initial Management of Kidney Stones." *American Family Physician* 64 (2001): 1329.

Pozzi, Claudio, et al. "Corticosteroids in IgA Nephropathy: A Randomized Controlled Trial." *The Lancet* 353 (1999): 883.

Roberts, Rosebud O., et al. "A Population-Based Study of Daily Nonsteroidal Anti-Inflammatory Drug Use and Prostate Cancer." *Mayo Clinic Proceedings* 77 (2002):219–225.

Rovner, Eric S., and Alan J. Wein. "The Treatment of Overactive Bladder in the Geriatric Patient." *Clinical Geriatrics* 10 (2002): 20–35.

Roy, Sushmita, et al. "Autosomal Recessive Polycystic Kidney Disease: Long-Term Outcome of Neonatal Survivors." *Pediatric Nephrology* 11 (1997): 302–306.

Rubin, Mark A., et al. "Alpha-Methylacyl Coenzyme A Racemase as a Tissue Biomarker for Prostate Cancer." *Journal of the American Medical Association* 287 (2002): 1662–1670.

Russo, Paul. "Evolving Understanding and Surgical Management of Renal Cortical Tumors." *Mayo Clinic Proceedings* 75 (2000): 1233–1235.

Saunders, Carol S. "Permanent Hair Dye Increases Risk of Bladder Cancer." *Patient-Care* 35 (2001): 10.

Savige, J., M. Buzza, and H. Dagher. "Haematuria in Asymptomatic Individuals." *British Medical Journal* 322 (2001): 942.

Schiffl, Helmut, et al. "Daily Hemodialysis and the Outcome of Acute Renal Failure." *The New England Journal of Medicine* 346 (2002): 305–310.

Seal, G. Mark. *The Patient's Guide to Urology*. Toledo, OH: High Oaks Publishing Company, 1995.

Sharp, Jennifer D., et al. "Bladder Cancer Detection with Urinary Survivin, An Inhibitor of Apoptosis." *Frontiers in Bioscience* 7 (2002): 36–41.

Shekarriz, Bijan, Hsueh-Fu Lu, and Marshall L. Stoller. "Correlation of Unilateral Urolithiasis With Sleep Posture." *Journal of Urology* 165 (2001): 1085–1087.

Smith, Shannon D., et al. "Urine Detection of Survivin and Diagnosis of Bladder Cancer." *Journal of the American Medical Association* 285 (2001): 324–328.

Sobel, Rachel. "A Tailor-Made Vaccine." *U.S. News & World Report* 128 (2000): 53.

Steineck, Gunnar, et al. "Quality of Life after Radical Prostatectomy or Watchful Waiting." *The New England Journal of Medicine* 347 (2002): 790–796.

Stenson, Jacqueline. "Annual PSA Testing May Be Unnecessary for Many Older Men." *Reuters Health* (May 21, 2002).

Stone, John H. "Wegener's Granulomatosis: More than a Vasculitis." *The Journal of Musculoskeletal Medicine* 18 (2001): 426.

Sutton, J.M. "Evaluation of Hematuria in Adults." *Journal of the American Medical Association* 263 (1990): 2475–2480.

Sweeney, William E., et al. "Treatment of Polycystic Kidney Disease with a Novel Tyrosine Kinase Inhibitor." *Kidney International* 57 (2000): 33–40.

Tanagho, Emil A., and Jack W. McAninch, eds. *Smith's General Urology*. Norwalk, CT: Appleton & Lange, 1995.

Terasaki, P.I., et al. "High Survival Rates of Kidney Transplants from Spousal and Living Unrelated Donors." *New England Journal of Medicine* 333 (1995): 333–336.

Trotto, Nancy E. "Contemporary Management of Bladder Cancer." *Patient Care* 34 (2000): 72.

Trotto, Nancy E. "Managing Kidney Cancer—A Malignancy on the Increase." *Patient Care* 34 (2000): 91.

U.S. Department of Health and Human Services, Health Resources and Services Administration, Office of Special Programs, Division of Transplantation, and United Network for Organ Sharing. "2000 Annual Report of the U.S. Scientific Registry for Transplant Recipients and the Organ Procurement and Transplantation Network: Transplant Data: 1990–1999." http://www.unos.org/Data/anrpt=_main.htm. Accessed February 11, 2002.

U.S. Renal Data System. "USRDS 2001 Annual Data Report: Atlas of End-Stage Renal Disease in the United States." Bethesda, MD: Author, 2001. http://www.usrds.org/. 2001. Accessed December 18, 2001.

U.S. Transplantation Data. United Network for Organ Sharing http://www.unas.org/data December 2003. Accessed December 17, 2003.

Vapnek, Jonathan. "Urinary Incontinence: Screening and Treatment of Urinary Dysfunction." *Geriatrics* 56 (2001): 25.

Ward, Michael M. "Changes in the Incidence of End-Stage Renal Disease Due to Lupus Nephritis, 1982–1995." *Archives of Internal Medicine* 160 (2000): 3136.

Wascher, Robert A. "Accuracy of Prostate-Specific Antigen (PSA) Test." Mensnewsdaily.com. http://mensnewsdaily.com/archive/w/wascher/03/wascher072703.htm. July 27, 2003. Accessed August 2003.

Wegener's Granulomatosis Association. "What Is Wegener's Granulomatosis?" http://www.wgassociation.org/aboutwg/. Accessed December 18, 2001.

Wershub, Leonard Paul. *Urology: From Antiquity to the 20th Century.* St. Louis, MO: Warren H. Green, 1970.

Williams, Rebecca D. "Living Day-to-Day with Kidney Dialysis: Quality Improvements Continue for Devices and Clinics." U.S. Food and Drug Administration. http://www.fda.gov/fdac/features/1998/198=_dial.html. 1998. Accessed February 4, 2002.

Wilt, Timothy J. "Treatment Options for Benign Prostatic Hyperplasia." *British Medical Journal* 324 (2002): 1047–1048.

Wish, J. B., and Kelly A. Weigel. "Management of Anemia in Chronic Kidney Disease (Predialysis) Patients: Nephrology Nursing Implications." *Nephrology Nursing Journal* 28 (2001): 341.

Woolhandler, S., et al. "Dipstick Urinalysis Screening of Asymptomatic Adults for Urinary Tract Disorders. I. Hematuria and Proteinuria." *Journal of the American Medical Association* 262 (1989): 1214–1219.

Xu, Jianfeng, et al. "Germline Mutations and Sequence Variants of the Macrophage Scavenger Receptor 1 Gene Are Associated with Prostate Cancer Risk." *Nature Genetics.* http://www.nature.com/cgi-taf/DynaPage.taf?file=/ng/journal/vaop/ncurrent/abs/ng994.html&filetype=&=_UserReference=C0A804EF4652784A4CA2A7923E753D8749C6. September 16, 2002.

Zoler, Mitchel L. "Erythropoietin Before Dialysis Linked to Prolonged Survival." *Internal Medicine News* 34 (2001): 2.

Index

About the Author

STEPHANIE WATSON is an independent scholar who has written and contributed to numerous works, including *World of Genetics* and *Science and Its Times*. She is coauthor of *The Endocrine System* in Greenwood's Human Body Systems series.